CULTURE AND CULTURAL ENTITIES

SYNTHESE LIBRARY

STUDIES IN EPISTEMOLOGY,

LOGIC, METHODOLOGY, AND PHILOSOPHY OF SCIENCE

VOLUME 170

CITY COLLEGE STUDIES
IN
THE HISTORY AND PHILOSOPHY OF SCIENCE AND TECHNOLOGY

JOSEPH MARGOLIS
Temple University

CULTURE AND CULTURAL ENTITIES

Toward a New Unity of Science

D. REIDEL PUBLISHING COMPANY
A MEMBER OF THE KLUWER  ACADEMIC PUBLISHERS GROUP
DORDRECHT / BOSTON / LANCASTER

Library of Congress Cataloging in Publication Data

CIP

Margolis, Joseph Zalman, 1924–
Culture and cultural entities.

(City College studies in the history and philosophy of science and technology) (Synthese library ; v. 170)
Includes index.
1. Philosophy—Addresses, essays, lectures. I. Title.
II. Series.
B29.M3673 1983 110 83–4635
ISBN 90–277–1574–2

Published by D. Reidel Publishing Company,
P.O. Box 17, 3300 AA Dordrecht, Holland.

Sold and distributed in the U.S.A. and Canada
by Kluwer Academic Publishers
190 Old Derby Street, Hingham, MA 02043, U.S.A.

In all other countries, sold and distributed
by Kluwer Academic Publishers Group,
P.O. Box 322, 3300 AH Dordrecht, Holland.

Printed in The Netherlands

For Marjorie Grene,
for the pleasure of her company

CITY COLLEGE STUDIES IN
THE HISTORY AND PHILOSOPHY OF SCIENCE AND TECHNOLOGY

EDITORS' PREFACE

Recent years have seen the emergence of several new approaches to the history and philosophy of science and technology. For one, what were perceived by many as separate, though perhaps related, fields of inquiry have come to be regarded by more and more scholars as a single discipline with different areas of emphasis. In this discipline any profound understanding so deeply intertwines history and philosophy that it might be said, to paraphrase Kant, that philosophy without history is empty and history without philosophy is blind.

Another contemporary trend in the history and philosophy of science and technology has been to bring together the English speaking and continental traditions in philosophy. The views of those who do analytic philosophy and the views of the hermeneutists have combined to influence the thinking of some philosophers in the English speaking world, and over the last decade that influence has been felt in the history and philosophy of science. There has also been a long standing influence in the West of continental thinkers working on problems in the philosophy of technology. This synthesis of two traditions has made for a richer fund of ideas and approaches that may change our conception of science and technology.

Still another trend that is in some way a combination of the previous two consists of the work of those characterized by some as the "Friends of Discovery" and by others as the bringers of the "New Fuzziness." This approach to the history and philosophy of science and technology concentrates on *change, progress*, and *discovery*. It has raised old epistemological questions under the guise of the problem of rationality in the sciences. Although this approach has its origins in the work of Thomas Kuhn in the United States, attempts to express his ideas in explicit set-theoretical or model theoretic terms are now centered in Germany.

The more traditional approaches to the history and philosophy of science and technology continue as well, and probably will continue as long as there are skillful practitioners such as Carl Hempel, Ernest Nagel, and their students.

Finally, there are still other approaches that address some of the technical problems arising when we try to provide an account of belief and of rational choice and these include efforts to provide logical frameworks within which we can make sense of these notions.

This series will attempt to bring together work from all of these approaches to the history and philosophy of science and technology in the belief that each has something to add to our understanding.

The volumes of this series have emerged either from lectures given by an author while serving as an honorary visiting professor at The City College of New York or from a conference sponsored by that institution. The City College Program in the History and Philosophy of Science and Technology oversees and directs these lectures and conferences with the financial aid of the Association for Philosophy of Science, Psychotherapy, and Ethics.

MARTIN TAMNY
RAPHAEL STERN

TABLE OF CONTENTS

PREFACE

I have tried to make a fresh beginning on the theory of cultural phenomena, largely from the perspectives of Anglo-American analytic philosophy. This is partly because of my own training; partly because I am persuaded that the analytic tradition can be enriched, without betraying its admirable sense of rigor, in such a way that it can incorporate important and subtle questions that it has characteristically ignored in recent decades; and partly because I believe that philosophy prospers by drawing its arguments out dialectically from the partial achievements of the strongest and most admired contributions of its immediate past. The themes developed I have approached in a variety of ways in a number of other publications: but here, I have tried to bring the discussion to bear on the converging issues of the so-called human studies – in particular, on the topics of language, history, action, and art. I have already, I may say, moved on in this direction in ways that look forward to a full rapprochement between Anglo-American and Continential philosophy. But I can hardly deny that I have been infected already with a sense of the importance of that.

The occasion for first preparing these essays was a series of lectures that I was invited to give at City College and at the Graduate Center, City University of New York, during the academic year 1979–1980, through the Program in the History and Philosophy of Science and Technology. I served at that time as Visiting Professor in the Department of Philosophy, City College; and the lectures and some related seminars were co-sponsored by The Conference on History and Philosophy of Logic, Science, and Technology. I particularly wish to thank Professor K. D. Irani for his kind invitation to join the Department in this capacity; to Professor Martin Tamny and Dr. Raphael Stern for their friendship and unfailing good will in launching and shepherding the entire venture; and to Professor Marshall Cohen for the invitation to present a good part of Chapter 3 at the Graduate Center. Parts and versions of various chapters were also presented at the German-American Colloquium on the Philosophy of Technology (1981), Lehigh University (1980), Michigan State University (1980), State University College at Brockport (1979), a joint meeting of the Fullerton Club and the Washington Philosophy Club (1979), the Fourth International Wittgenstein Symposium (1979), the International

Society for the Comparative Study of Civilizations (1979), the Center for Philosophy of Science and Department of History and Philosophy of Science, University of Pittsburgh (1978), the New Jersey Regional Association for Philosophy (1978), and the American Association for the Advancement of Science (1976). I hadn't quite realized how many trials of these sorts were actually involved in shaping these essays; but I am enormously grateful to the host institutions and colleagues for the opportunity to test and refine my views.

I must also thank Mrs. Grace Stuart, who prepared the manuscript in her unfailingly splendid way. These essays have been somewhat delayed by a longish illness, which fortunately is now past. But the interval has enabled me to benefit from additional reflections undertaken with an entirely new project in mind.

Philadelphia, Pennsylvania J. M.
November 1982

ACKNOWLEDGEMENTS

With alterations and additions, some rather extended, (1) has appeared as 'Nature, Culture, and Persons' in *Theory and Decision*, XIII (1981); (2) as 'The Concept of Consciousness', *Philosophic Exchange*, III (1980); (4) as 'Action and Causality', *Philosophical Forum*, XI (1979); and (5) as 'Puzzles about the Causal Explanation of Human Actions', in Larry Laudan (ed.), *Mind and Medicine: Problems of Explanation and Evaluation in Psychiatry and the Bio-medical Sciences* (Berkeley: University of California Press, 1983). Permission to reprint has been granted.

1

NATURE, CULTURE, AND PERSONS

One cannot, pursuing in a disciplined way any aspect of the comparison of cultures, fail to sense a certain uneasiness about the prospects and appeal of the unity of science. In the Anglo-American philosophical tradition, the program of unity has been almost hopelessly linked with the fortunes of a defeated positivism and the striking lack of success of all forms of reductive materialism.[1] In the Continental tradition, notably in the efforts of philosophical hermeneutics, the contrast between *Naturwissenschaften* and *Geisteswissenschaften* has, almost since Dilthey's original speculations, conveyed a strongly dualistic conviction.[2] Still, the need for a unity of science is conceptually irresistible. It serves as the methodological counterpart of the abiding intuition of the integrity and conceptual coherence of human existence. For contingent reaons, its actual pursuit is taken as flat evidence of a reductive intent. One sees this, for example, symptomatically, in the appearance of a new, explicit dualism within analytic philosophy of science – as in the most recent speculations of Karl Popper regarding what he calls World 3.[3] Roughly speaking, the conceptual requirements entailed in reconciling our picture of physical nature and human culture, *at the present stage of philosophical work*, cannot but reflect the continued repudiation of ontic dualism and the recognition of the inadequacy, even within the physical sciences, of the explanatory models that inevitably favored reductionism.[4]

What we need, short of a completely reformulated unity, is a proper grasp of the details of a generous philosophical strategy by which the extremes of dualism and reductionism can be effectively avoided – at the same time the admitted distinctions of the *Geisteswissenschaften* are preserved. In that sense, the program required is both ontological and methodological. What I mean is simply that we need to understand what must be included in a theory of persons and in a theory of explanation in order to make the best sense of such an effort as that of comparing distinct civilizations. In order to vindicate the rigor and validity of our empirical studies, therefore, we must at some point step back to reflect on their conceptual underpinnings. At the present time, the most promising ontological orientation can only be what may be called nonreductive materialism.

Here, a small distinction is needed. If we contrast ontic dualism with attribute dualism – in the sense in which we contrast a dualism of substances (in effect, Cartesianism) and the irreducibility of various kinds of attributes (in particular, culturally significant attributes) to purely physical attributes – then, by a nonreductive materialism we may understand a form of attribute dualism.[5] Ontic dualism entails attribute dualism, but not *vice versa*. It is the irreducibility of various kinds of attributes to attributes of a purely physical sort that (at the present stage of philosophical reflection) is the essential concern of the mind/body problem – not the thesis that real entities (of the relevant kinds) are all and only composed of what physical objects are composed of. The interesting thing is that *nearly all* relevant ontological theories, reductive as well as nonreductive, in effect admit attribute dualism. This may seem surprising. But the reason the thesis is allowed even by the most extreme reductionists is that the concession is thought to be benign. Thus, for instance, Wilfrid Sellars admits the irreducibility of the intentional. He wrongly believes that it does not affect reductive programs because he assimilates the intentional to merely formal, logical, or semantic functions or "roles" that may be "added to", or imposed upon, an otherwise adequate reductive account of the real world.[6] But if one understands the intentional in terms of the actual psychological or internal capacities of human persons and of what they actually produce and do in the setting of a culture (for example, actually explain physical phenomena), then it is reasonably clear that Sellars has defeated himself out of his own mouth – and, in effect, identified the insuperable difficulty confronting all forms of reductive materialism.[7] For, the irreducibility of the intentional counts, on any plausible analysis, as the irreducibility of actual human powers and products; and that forces us either to ontic or attribute dualism. In the analytic literature, attribute dualism – in particular, a dualism involving intentionality as a real property – is tantamount to the admission of emergence under conditions incompatible with the traditional program of the unity of science.[8]

Of course, to speak of "dualism" in the double sense signified is merely rhetorical economy, but it is an economy that bears directly on the conceptual strategies that must be considered. Canonically, dualism maintains that there are fundamentally different substances, of which real or natural things are composed: Cartesian dualism is the paradigm. Attribute dualism is not so much a dualism as the insistence or acknowledgement that the real properties of things are not reducible to being all and only physical properties or properties of some other single kinds, taken in a similar and suitably fundamental sense. Nevertheless, insofar as the ascription of properties to things must be adequated to the nature of the things thus qualified, a rich version of "attribute dualism" (for instance, Descartes's own) inevitably poses the threat of ontic dualism.

We see, therefore, the profound interconnectedness of methodological and ontological concerns bearing on the explanation of human cultures. But, for the moment, we may content ourselves with emphasizing that the most viable form of attribute dualism cannot but be a form of materialism. Though this is not logically required, any review of the ontological and methodological scandals of the history of ontic dualism, as well as of the achievements of the physical and biological sciences, leaves no doubt about the importance of reconciling cultural emergence with a form of ontic monism – *a fortiori*, materialism. This is not to say that the nature of the physical world is adequately understood. It is to say only that whatever proves to be an adequate account will have to serve as the substantive foundation for the emergence of culture as well, *and* that a kind of Leibnizian panpsychism is either vacuous or false.[9]

The general concessions on the methodological side are just as easy if not easier to specify, since the inadequacies of the hypothetico-deductive model of explanation have already forced themselves on all engaged in resolving the philosophical puzzles of science, in terms of the attempt to account for the explanation of purely physical phenomena. The principal difficulties are quite straightforward. The admission of statistical or probabilistic laws entails that explanatory models in science need not be deductive in nature and that laws need not be universal in form. Where, as in quantum physics, statistical laws are thought to reflect objective tendencies in the natural world, not mere cognitive limitations on the part of human investigators, determinism – at least ubiquitous determinism – is undermined.[10] What is left seriously unclear is the nature of causality itself. For, for one thing, it is a commonplace of conservative theories of causality (characteristically, though not necessarily, linked with the prospects of reductionism) that causal occurrences, even if observed in singular instances, presuppose covering laws that are universal in form.[11] It is the universal form of the law that insures the deductive model; and it is the presumed adequacy of the laws of the physical sciences – so qualified – that insures both reductionism and determinism. The admission, as covering laws, of probabilistic generalizations of a *causal nature* may therefore seem an anomaly to a Humean or to an advocate of the traditional unity of science program.[12] There are, however, attempts to extend an adjusted version of the hypothetico-deductive model to cover cases in which the explananda are statistical laws; these require that the explanans itself contain at least one sentence designating a statistical law.[13] But the general adequacy of the model nevertheless remains in doubt. The point is not so much to insure a conceptual foundation for human freedom – though it is clear that human freedom is incompatible with a thoroughgoing determinism that treats the actual and the (naturally) possible as coextensive; it is rather that the methodologically fundamental notions of explanation, natural law, and causality are completely upset by the admission of probabilistic laws, on either epistemological or metaphysical grounds (but more significantly, on the latter). And that means that the very conception of the unity of science must be recovered in a radically new way – that obviates both reductionism and (hard) determinism. Hence, the inherent tendencies of the philosophy of the physical sciences cannot fail to be hospitable to the special features of the *Geisteswissenschaften*. In fact, broadly speaking, probabilism demonstrates that there is no conceptual connection between causality and nomic universals and that if causality is to be preserved at all, it must be linked rather more informally to models of agency (applicable in singular cases without presuming universal covering laws or even covering laws themselves), attenuated in a systematic way from the paradigms of purposive human behavior to instances of animal behavior to instances of inanimate agency, under the constraint of a favored explanatory theory.[14] Otherwise, there would be absolutely no way of distinguishing between lawlike and nonlawlike regularities. (We shall return to agency shortly.)

When it was held that nomic regularities had to be universal in form, it was supposed that the difference involved somehow had to do with the counterfactual entailments of a genuine law;[15] but the admission of probabilistic laws eliminates (or at least threatens) a univocal reference to counterfactuality in eliminating (or threatening) universality. For, that particular instances of a set of things to be explained might fail, given objective tendencies, to manifest what the law favors only probabilistically, cannot for that reason alone be construed counterfactually; propensity (even frequency) considerations may, however, recover the counterfactual.[16] It is also apparent that the logical constraints often imposed on so-called genuine laws of nature – for example, that they or their *ceteris paribus* conditions make no essential reference to the earth (Mendel's laws) or to the spatiotemporal peculiarities of our solar system or galaxy or universe (the possible contingency of the "big bang") – run the risk of precluding genuine laws altogether.[17] It would hardly do, obviously, to permit the very concept of causality to depend on such fortunes; conceivably, terrestrial and galactic references may be construed as not restrictive in that sense. But on the most stringent reading

– J. J. C. Smart's for instance – social and psychological laws are quite impossible. Still, the principal conceptual difficulty confronting the hypothetico-deductive model remains the one posed by probabilistic laws; for, although, in the physical sciences, it may be argued (it may be contested, also) that relevant probabilized regularities tend toward some invariance as a limit, it seems quite unlikely that a similar claim can be favored or favored in a comparably strong way in the human sciences, particularly if reductive materialism is denied or put at risk.

In the sense here sketched, the prospects of recovering an adequate model of the unity of science pretty well require the rejection of what was supposed to obtain in the traditional version of that unity. To give up reductionism and a deductive model of explanation is, in effect, to be hospitable to alternative forms of emergence; the question of *what* kind of causal account would be methodologically appropriate for different disciplines cannot then be antecedently decided for all disciplines, cannot but be conceptually dependent on the ontological pecularities of the particular domain investigated. For example, how a scientist's grasp of the history of his own discipline causally affects his own work can, fruitfully or nonvacuously, hardly be said to function (for explanatory purposes) in the same way in which biochemical changes at the level of the DNA code function. Also, the admission of emergence in this sense inevitably introduces the problem of constructing a single paradigm of explanation – where, that is, admissible explanations may well require different sorts of interlevel reference: as, for instance, in biology, where changes in the subsystems of organisms causally depend on factors specified both at the genic level and at the level of molar organisms responding to environmental forces; or again, as in psychology, where changes in behavior causally related to genetic factors, contingently affected anatomical and physiological factors, and culturally induced learning, obtain.[18] Failure to gain an ontological reduction (or at least a fair sense of the way in which such a reduction could be accomplished) forces us, in the context of our methodological concessions, to admit as well the deep informality that may infect causal explanations in an essential way.

From the point of view of the human and cultural studies, the crucial conceptual adjustments – at once methodological and ontological – concern the provision of an adequate account of causality among cultural phenomena and the viability of an explanatory model in which *emergent* human agency, though compatible with nomological regularities (of either an invariant or probabilistic sort) operating at least at the purely physical level (at which actions may be "embodied"), may not itself fall under covering laws at all.

(The topic of embodiment is taken up, below.) If, on the foregoing considerations, causality is not conceptually linked to nomic universals, and if singular causal connections may be discriminated without knowledge of any lawlike regularity, then – regarding the explanation of historically and culturally distinctive (even unique) phenomena – it may well be necessary to develop a model that admits causality without covering laws of any sort.[19] This seemingly heterodox proposal is, on the face of it, not merely coherent but actually promising, given the logically uncertain status of natural laws, the difficulty of formulating a uniform test of counterfactuality, the independence of causality and nomologicality, *and*, of course, the peculiar requirements of the cultural disciplines themselves.

It is precisely at this point that the distinctive themes of the hermeneutic tradition bearing on the contrast between the natural and the cultural disciplines begin to look recoverable within a framework in which attribute, but not ontic, dualism is tolerated and in which one searches for a resiliant conception of the unity of science. For example, the provision of such an explanatory model is promisingly congruent with the more or less anecdotal form of causal explanation (which is itself not necessarily meant to preclude the relevance of probabilistic laws, if there are any, of psychological development or cultural transmission or the like) favored, say by Freud,[20] or Max Weber,[21] or Geertz,[22] or Goffman,[23] or indeed in any of the usual specimens of historical explanation. The inconclusiveness, for instance, of Hempel's well-known attack on William Dray's conception of history, and the absence of an alternative and ramified causal model in Dray's account, testify to the fresh prospects that the present line of reasoning makes possible.[24]

Here, again, methodological and ontological distinctions prove to be mutually dependent. The general problem of the ontology of culture and of culturally emergent entities concerns how to explain the nature of a system in which: (i) ontic dualism is rejected; (ii) attribute dualism is admitted; (iii) reductionism is rejected; (iv) causality is construed as logically independent of, but compatible with, nomologicality; and (v) causality obtains in an interlevel way to include the culturally emergent.

Two distinctions serve to complete our sketch and to suggest the way in which to resolve the puzzles of the new unity of science. One concerns the analysis of a cultural attribute; the other, a theory of the most generic ontic features of culturally emergent entities – in particular, persons, but also artworks and sentences, and, by extension, even conventional behavior. The key theme must be that persons – minimally, human persons – are culturally emergent entities distinguished primarily by their linguistic ability and whatever such ability makes possible. Among humans, we must think of the cultural training (however achieved) of a set of unusually talented animals, though we need not foreclose on the possibility of artificial persons.[25] Furthermore, human beings, like other sentient animals, exhibit mental attributes pre-culturally, and (in the human case) also sub-culturally even within the interval in which they emerge as persons. That is, the psychological features of human beings obviously cannot be accounted for exclusively in cultural terms, and they form a natural endowment on which the cultural itself depends. Also, the very existence of other culturally emergent entities and phenomena – artworks, language, conventional behavior – causally depends on the existence and distinctive activity of human persons, though those phenomena (excluding behavior) characteristically lack any mental or psychological attributes. This suggests that an essential distinction is required to mark the contrast between the merely psychological and the cultural – which, it should be noted, was the heart of Dilthey's insistence on not conflating the psychological and the hermeneutic.[26] It may be admitted that Chomsky's thesis of linguistic innatism complicates the picture of persons as culturally emergent entities.[27] But, for one thing, Chomsky himself concedes that every natural language must be learned in a cultural setting – say, French or some Eskimo tongue. And secondly, Chomsky himself offers absolutely no explanation of what it means to say that linguistic rules as such *could* be part of our pre-cultural, natural endowment. The notion, thus pressed, may well be incoherent, as the elaboration of cultural attributes makes clear. And, in fact, the difficulty with Chomsky's account in this regard may well infect other versions of structuralism – notably, Lévi-Strauss's[28] – though not necessarily (at least for the reasons in question) those structuralist or quasi-structuralist accounts empirically developed and explicitly opposed to any form of apriorism – for example (in rather different ways) Piaget's[29] – or Marx's or Freud's.

Now, it turns out that, although the psychological and the cultural are distinct, the characterization of mental states – whether of humans or of nonhuman animals – is, where such states are cognitively significant, modelled (not merely described) in linguistic terms. That is, all cognitive states are assigned (in various ways) propositionally formed "objects." Since human beings are able to perform linguistically, this seems perfectly natural, even where (as with thoughts) we may be prepared to concede nonlinguistic, or nonlinguistically informed, mental states. In the case of nonhuman aninals, to the extent that we attribute cognition at all (as in the perception of dogs, or, more dramatically, in the mastery of a fraction of an alien natural language by chimpanzees and gorillas), we must treat linguistic modelling as purely heuristic, however unavoidable. In that sense, animal psychology (short of the reductive programs of radical behaviorism)[30] cannot but be anthropomorphized.[31] This obviously complicates the attempt to distinguish between the psychological and the cultural. Generally speaking, however, the distinction of the psychological (not the cultural) follows fairly closely the original work of Franz Brentano, though with considerable refinement and adjustment of his particular theses, for instance along the lines provided by Edmund Husserl.[32] At this point, however, it is strategically more important to make systematic provision for a rich conception of the psychological and the cultural within causal and explanatory contexts then to attempt a fully satisfactory account of those notions themselves. On Brentano's view, the psychological is distinguished by its *intentionality*, that is, that mental states are "about" something, or "directed upon" an object, whether or not the object is real or actual and whether or not what one (in a cognitively relevant sense) supposes (propositionally) about that object is true. For example, Tom may be thinking about a unicorn, or Tom may believe that he could capture one. In the first instance, the object is nonpropositional; but since the state in question is a cognitively pertinent one, a propositional object obtains in principle as well (though we may not know what it is). Since the only way to model propositional content is linguistically, we cannot avoid the heuristic dependence already noted.

On the other hand, words and sentences are said to be meaningful, or to have meaning, in accord – in some sense extraordinarily difficult to specify – with the institutionalized and historically shifting life of some society of actual human persons. In this respect, languages are said to exhibit inten*s*ionality, by which one means to signify a certain logical peculiarity of linguistic elements, *not* a feature as such of psychological states. The peculiarity is at least partially captured by the following condition: that, in certain linguistic contexts, one cannot substitute codesignative terms, preserving truth. For example, if Tom believes that Cicero denounced Catiline, then, though Cicero and Tully are one and the same person, it does not follow that Tom believes that Tully denounced Catiline. There are a great many circumstances in which cognate difficulties arise – most important, perhaps, where terms are admitted to be coextensive but still differ in sense (for instance, assuming the context is clear, "the one and only grandson of A" and "husband of B"). Inten*t*ionality, in Brentano's sense, was meant to secure the contrast between the psychological and the physical; on the other hand, inten*s*ionality, in the current lingistic or logical sense, is opposed to the so-called extensional (which concerns designated equivalences regarding sets of things in a given domain). The trouble is that Brentano wrongly supposed that the inten*s*ional was also a sufficient mark of the inten*t*ional (in the psychological sense). And so the two are often wrongly taken to be the same or coextensive notions. But they are not.[33] However, since there is no language unless there are (in the timeless sense) speakers of that language, and since language users must possess psychological states, a subset at least of the psychological states of human persons are characterizable in both inten*t*ional and inten*s*ional terms. In particular, all linguistically informed mental states (as in thinking subvocally, or thinking what to say) must be characterized in a novel way that is jointly inten*t*ional and inten*s*ional.

The importance of this complication cannot be overestimated. For one thing, it is just this sort of intentionality that Sellars had in mind in conceding the nonreducibility of the intentional. Since the linguistic seemed to be purely formal in essentially the same sense in which the logical is, Sellars wrongly supposed that the admission of the intentional would not be inimical to reductive materialism. But as soon as we grant that there is a set of actual psychological powers that persons exhibit, that involves in a complex way both the inten*t*ional and the inten*s*ional, we cannot fail to see that the admission of language is the paradigm both of what is *culturally* emergent *and* of what is causally efficacious within the cultural domain. Alternatively put, reference to language is, equivocally, reference to actual speech or to an abstract system of sentences. Confusion between the two (as in Sellars) contributes toward deflecting us from the implications of real linguistic abilities. A physical utterance may exert a causal force; but an utterance that has a certain propositional import may affect a being capable of grasping that import in a way altogether unlike the way in which a physical force operates. Only a reductive conviction could resist this concession. Moreover, if the distinctive intentionality of speech behavior is irreducible (we may, if we wish, write "intentional" in the linguistic or cultural sense, with a capital I), then we are absolutely obliged to consider a whole range of different causal relations, among which, merely grasping the content of a message may play a distinctive causal role. This, then, is the clue to recovering, within a causal model, Dilthey's fundamental distinction. Furthermore, since language behaves inten*s*ionally, and since at least the reductive ideal of discourse restricted to the purely physical is supposed to behave extensionally, it is impossible to understand Chomsky's thesis that there are rules of deep grammar that we are innately endowed with – that, on the evidence (and increasingly, on Chomsky's own speculations), cannot be applied (if we admit that they can be applied at all) without a knowledge of the semantic and even nonlinguistic, experiential constraints within which they actually function. For, there is no viable sense in which those semantic and nonlinguistic constraints could be provided innately, or transmitted genetically, in some species-specific way; and there is no clear sense in which *rules* – which must themselves be construed intensionally (and which Chomsky explicitly opposes to invariant natural laws and transcendental conceptual constraints) – could be genetically transmitted.

Perhaps the required distinctions may be put in the following way. As a purely abstract system of sentences having semantic and referential import, language may be said to exhibit inten*s*ionality, but in a sense (congruent with Sellar's position) entirely neutral to the analysis of real psychological properties; on the other hand, creatures that manifest real psychological states, whether linguistically informed or not, behave inten*t*ionally (more or less in Brentano's sense). However, only a creature that has mastered language and behaves in a way informed by such mastery acts in a culturally significant way. Such a creature behaves, we may say (invoking a term of art), Intentionally; his real psychological states cannot be correctly described except in terms of intentional properties (i) that are now intensionally qualified and (ii) that may not be always or entirely accessible to him cognitively or computationally. So the Intentional – in the cultural sense – is distinctive in two important respects. First of all, since language, construed as actual speech, constitutes a real psychological ability (the necessary condition of all cultural phenomena and the most fundamental manifestation of the cultural), certain of the real psychological attributes of human beings must be characterized in intensional ways. Think, for instance, of identifying what is done as an insult or as a greeting or a signal. Secondly, given the complexity of the natural language and local culture in which an entire society is groomed, no particular individual could have internalized, in an infra-psychological or cognitive sense, the entire range of cultural or linguistic traditions and rules that contingently influence and inform his actual behavior and mental states. Hence, the states and behavior of human persons manifest intentional properties, features of "aboutness", that are both informed by a society's traditions and practices and are generated only partly by means of such persons' internal cognitive or computational abilities. This helps to contrast the so-called hermeneutic and (purely) psychological uses of "intentional"; at the same time, it raises a fundamental objection to the new cognitivism.[34]

The cultural itself is distinctive, then, in that the linguistically informed behavior and psychological states of human agents – not merely abstract sentences – exhibit intensionally qualified intentional features; and the effects or products of the activities of such agents – whether generated in a fully calculative way or at least causally subject (in part) to the operative practices and traditions of a society that they could not have completely internalized – also exhibit intentional features that are intensionally qualified or are open to alternative interpretation. Think, for instance, of linguistic utterances, rituals, and works of art.[35] The Intentional, in the cultural sense, cannot be restricted to the infra-psychological, but it still requires real ascriptions of the intensional to the internal states of human beings. This is not true, for instance, of the higher animals lacking language, in spite of the fact that they are capable of cognitive states.[36] A suggestive way of marking the *sui generis* nature of the cultural is this: the distinctive behavior of human beings may be said to be *cognigenic* more than completely or always cognitive, in the sense both that cognition occurs among animals below the level of linguistic ability and in the sense that the cognitively focused behavior and mental states of human beings are influenced by the effective practices of that living society in which those agents become culturally effective themselves; and the distinctive activity or products or practices yielded by human labor may be said to be *cognignomic*, in the sense both that they are produced by culturally informed agents and are recognized by them as inviting Intentional interpretation.[37] Hence, the cultural cannot be reduced to the physical unless (against the usual admission of reductionists themselves) the various forms of intentional phenomena can be suitably reduced.

We must, of course, take a very cautious view of the claim that human languages are rule-governed.[38] The fact is that, although human language is distinctly rule-*like*, there is no reliable empirical evidence that there is a set of explicitly formulable rules that fit all the phenomena of intelligible speech *and* that are in some sense actually *used* by the speakers of a language. One can see the point very easily by considering that the meaning of any human utterance must depend on the extralinguistic context in which the utterance is produced, on background information and beliefs about the world that speaker and hearer separately and jointly possess.[39] And, if this is so for language, how much more certainly is it true, say, of period styles in dress and manners and painting and the like. So to say that language is rule-*like* is to say that there is a certain *intensional regularity* that informs the practice of speech, that cannot be completely formalized, that depends on context, and that is open to deliberate and accidental change over time. But *if* this is so, then we see at once why it is important, if the linguistic and the cultural are causally efficacious, that the causal be admitted to be conceptually independent of the nomological. There are no determinate universal rules (governing either language or other cultural phenomena) that linguistic ability presupposes; but there are more or less prevailing regularities that rough-and-ready rules of thumb or maxims (or, more ambitiously, institutions, traditions, genres, styles, and the like) reasonably collect; and *these* may inform the causal agency of human persons. They cannot however be distinguished on purely physical grounds; and what we may call their functional regularity cannot but be approximate only. Alternatively put, cultural practices and institutions exhibit an intensional regularity because: (a) the actual members of a society are groomed, before they are fully cognitively competent, in the historically developing practices of that society, to which, (b) their own improvisations may, both prospectively and retrospectively, be construed as continuations, confirmations, or extensions of those practices, which (c) are themselves always open to plural interpretation and plural lines of extension. What counts as an admissible continuation of a cultural practice can only be what, in some consensual sense, is in effect acknowledged by the members of a society confronting, contextually, new circumstances that invite their improvisation. This holds as well for conversation as for marriage practices, the giving of gifts, the handling of insults, dress, and the like. The regularities of cultural life, then, are not generated by an infra-psychological, cognitive grasp of explicable rules; but they can, retrospectively or perhaps for short predictive intervals, be conveniently approximated in terms of alternative, variably plausible rules. In fact, these last remarks may, not unfairly, be construed as an explication of Wittgenstein's famous notion of "forms of life".

In short, *if* cultural forces have a causal role, then it is impossible to deny that they may be subsumable only under what may be called "covering institutions" rather than covering laws – meaning by that, that a family of intensionally informed actions may conform loosely with some rule-like regularity that itself cannot be captured by any extensional invariance or by any merely statistical replacement of such an invariance. We may say, then, by way of summary, that the cultural is the Intentional – not in the sense either of the merely psychological or the semantic, but in the sense in which a subset of the psychological states of human persons and a subset of their actions and utterances are informed by rule-like regularities (assignable societally, never completely internalized infra-psychologically) that change (historically) under the causal influence of behavior similarly informed. So the cultural is not restricted to the linguistic, but is both informed by the linguistic and causally produced by creatures that can perform linguistically. Accordingly, human history signifies the diachronic change and replacement of rule-like regularities, or the occurrence (and the causes of the occurrence) of events intelligible only in terms of the culturally Intentional. So far, then, given the distinction of language and culture, we see that the admission of attribute dualism is both inevitable and quite independent of ontic dualism. In that sense, it is entirely hospitable to the project of recovering the unity of science.

But we must look a little more closely at the ontology of persons and of other cultural entities that, causally, depend on the agency of persons. There are, I believe, two generic, jointly necessary and sufficient, conditions that cultural entities satisfy. Both are critically tied to the concept of the Intentional just developed. Cultural entities, I should say, are embodied and are tokens-of-types. I believe that all and only cultural entities exhibit these features. They are, however, intended as distinctions of art. Think of the following relationships: Winston Churchill and his body, Michelangelo's *Pietà* and the marble it was cut from, the word "good" and the ink marks by means of which it is printed. On the argument intended, of each pair, what are related are not identical with one another, and the first is not composed of the second; nevertheless, the first could not exist without the second's existing or (in a sense to be supplied) without some suitable alternative existing. Furthermore, the first possesses some of the properties of the second as well as Intentional properties of the cultural sort – in virtue of possessing which it has emerged in a cultural context and there exists. That the existence of the first depends on the existence of the second illustrates the relationship I call *embodiment*.[40] It is a *sui generis* relationship that holds only of cultural entities and that accommodates both the irreducibility of the Intentional and the reality of persons and what they make and do. Also, it is clearly not a biological relationship, though, as in the case of persons, it may be manifested biologically.[41] So persons are embodied in their bodies (if there were any, artificial persons would perhaps be embodied in electronic gear); sculptures are embodied in stone or wood, for example; and words and sentences are embodied in ink marks and noises. Rule-like regularities either of language or of what presupposes linguistic ability, distinguish all such entities: roughly, in the case of persons, their rule-following abilities; in the case of what they produce and do, the rule-governed features of artifacts and behavior. But, as already remarked, rules themselves are hardly more than synchronic idealizations of relatively informal, shifting Intentional regularities that cultural entities exhibit. Also, conformity with rule-like regularities suggests (correctly) a considerable tolerance (in cultural contexts) of variant – even incompletely predictable – physical embodiments of given particulars. Think, for instance, of a Hungarian with a cold pronouncing the word "good" in an atmosphere in which smog affects auditory reception in a peculiarly dampened way.

Cultural entities are, also, tokens-of-types.[42] Types are heuristically introduced abstract particulars (like a Beethoven sonata, or the word "good"), solely intended for purposes of individuating tokens as instances of one and same type (alternative performances of the sonata, or alternative printings and utterings of the word). There *are* no types, and "token" is an ellipsis for "token-of-a-type". Embodied particulars are individuated as tokens-of-types. The token utterance "good" is embodied in a set of ink marks and is a particular instance of the type word "good". It is incoherent to compare types and tokens; only alternative tokens of the same or different types may be compared. Also, the token/type distinction is different from, and not reducible to, the member/set or the instance/kind distinction. The salient considerations include these: that, heuristically conceived, types are created and destroyed (unlike universals), and artists and speakers produce particulars (not sets). Otherwise, for instance, we should not be able to say that Dürer created *Melancholia I*, or we should be obliged to say that Beethoven created all the performances of his *Fifth Symphony*.[43] Parallel arguments hold for persons and words. Cultural entities, then, exist only as embodied tokens of a given type. By attending to the Intentional design or regularity of a given type, various physically different phenomena may be construed as embodying particular instances of the same type. There are, doubtless, a great many conceptual quarrels that would have to be met before the full advantage of these distinctions could be gauged; but they are not of immediate concern here.

What is crucial, rather, is that we possess an ontology of cultural entities that sufficiently displays the sense in which: (a) cultural entities are real; (b) emergent in an irreducible way; (c) compatible with the materialism that informs the physical and biological sciences; (d) capable of entering into causal relations; and (e) capable of supporting causal explanations. Such an ontology, then, provides the basis for the recovery of the unity of science – but in a way that respects the diversity of the sciences themselves. In particular, the new unity of science must concede the logical independence of causality and nomologicality; hence, for different sciences (say, portions of physics, physiological psychology, and economics), that causal relations may be subsumed under invariant or probabilistic laws or merely under covering institutions (that are themselves subject to causal change); *and* that, for different sciences, the criteria for designating an actual causal relationship are bound to vary widely. Without a doubt, the most drastic revision that the cultural disciplines impose on the unity of science (apart from the conceptual independence of causality and nomologicality) is that, contrary to the conventional view of the physical sciences, cultural causes can only be identified *intensionally*.[44] Still, this is not a disaster; for, since the cultural is materially embodied, once any cultural factor is identified, whatever advantages accrue to the extensional identification of purely physical phenomena may be assigned to it as well. For example, the *Pietà* exists only in the Intentional world of cultural life; but, since it is embodied in a block of marble, its identity and the causal role it plays may be traced in terms of its marble embodiment. The one, of course, remains different from the other.

We may draw one final distinction. Only cultural entities have histories. Physical objects persist through moderate changes,[45] and the causal explanation of their present properties may well depend (as with the fatigue of metals) on the ordered sequence of their spatiotemporal phases. But a history, in the relevant sense, involves *as well* the ordered sequence of the Intentionally qualified phases of an embodied entity. Hence, since the Intentional attributes and regularities that any particular cultural entity may be said to exhibit can be assigned only approximately, variably in accord with one's idealization of the Intentional uniformities of a culture, and under conditions of actual historical change affecting the very ascriptions made – in effect, considerations central to the hermeneutic puzzle itself[46] – the admission of cultural phenomena inevitably obliges us to concede the coherence at least of a moderate relativism at both the interpretive level (at which the Intentional import of a given phenomenon is assigned) and at the causal level (at which such a phenomenon so interpreted may affect other embodied *and* embodying phenomena). It is in this sense, for instance, that the way in which a scientist understands the history of his discipline actually affects, through a series of experiments perhaps, the physical disposition of a portion of the world.

Cultural entities, then, are *embodied* in the sense provided; and their properties include at least (what we may now call) *incarnate* properties, that is, properties adequate to their being entities of the (emergent) kinds they are. Both persons and artworks, say, being Intentional systems, have Intentional properties: just as they are embodied entities, their properties are incarnated or naturalized in appropriate physical or biological properties. The *Pietà*, for instance, does not merely have a representational function *associated*, in some way, *with* the physical properties of the marble in which it is composed; it *has* an (emergent) representational property in just the sense in which it exists as a sculpture. It cannot *have* that property except insofar as the property itself is indissolubly unitary. Similarly, persons *have* linguistic capacities, and these are emergent properties that they, as culturally emergent entities, actually have. So "incarnate properties" are specified as such in order to accommodate various forms of attribute emergence with respect to emergent entities; that is, there must be such properties in order to concede attribute dualism without conceding ontic dualism – relative to mental, functional, and Intentional emergence. Incarnate properties form a mixed collection, of course: they include at least the psychological capacities of linguistically competent humans, the culturally distinctive properties of artworks (characteristically neither mental nor living) produced by encultured humans, *and*, also, the psychological capacities of languageless animals, and (if not reducible) the biologically functional (but nonpsychological) properties of living systems. Nonliving machines, to which we ascribe some form of artificial intelligence, are (culturally) embodied systems; hence, they, too have incarnate properties, not merely "functional" properties, in that purely abstract sense in which the machine analogue of Cartesian dualism obtains.[47] Clearly, incarnate properties (of different kinds) are manifested by both embodied and emergent (but nonembodied) entities.

The point of the foregoing distinctions, however, is not to decide prematurely in favor of, or against, say, relativism in historical explanation. It is, rather, to fix the conceptual distinctions – both ontological and methodological – presupposed in all the central quarrels regarding the cultural disciplines (for instance, as in the comparative study of civilizations). To have shown that these provide for alternative theories that are at once conceptually quite different from theories regarding physical phenomena, coherent in themselves, compatible with fair constraints derived from our success in the physical sciences, is to have moved (I should like to think) very far in the direction of recovering the unity of science without the prejudice of its own history.[48]

NOTES

[1] For a general overview of the program and an impression of the original undertaking, see Robert L. Causey, *The Unity of Science* (Dordrecht: D. Reidel, 1977); Otto Neurath *et al.* (eds.), *International Encyclopedia of Unified Science*, Vols. I–II (Chicago: University of Chicago Press, 1938); and Paul Oppenheim and Hilary Putnam, 'Unity of Science

as a Working Hypothese', in Herbert Feigl *et al.* (eds.), *Minnesota Studies in the Philosophy of Science*, Vol. 2 (Minneapolis: University of Minnesota Press, 1958).

[2] See Wilhelm Dilthey, *Meaning of History*, H. P. Rickman (trans. and ed.) (London: Allen and Unwin, 1961); and Hans-Georg Gadamer, *Truth and Method*, trans. (from 2nd edn.), Garrett Barden and John Cumming (New York: Seabury Press, 1975); also, Peter Winch, *The Idea of a Social Science* (London: Routledge and Kegan Paul, 1958). Wittgenstein, whom Winch professes to interpret, may fairly be said to have tried to reconcile the apparently dualistic strain in hermeneutic thought – some believe, in too reductive a manner (*almost* "behaviorizing" mental states); but Winch remains noticeably dualistic, even within the analytic tradition.

[3] See Karl R. Popper and John C. Eccles, *The Self and Its Brain: An Argument for Interactionism* (New York: Springer International, 1977).

[4] See C. G. Hempel, *Aspects of Scientific Explanation* (New York: Free Press, 1965); and Oppenheim and Putnam, *loc. cit.*

[5] The full development of this thesis appears in Joseph Margolis, *Persons and Minds: The Prospects of Nonreductive Materialism* (Dordrecht: D. Reidel, 1978).

[6] Wilfrid Sellars, 'Philosophy and the Scientific Image of Man', in *Science, Perception, and Reality* (London: Routledge and Kegan Paul, 1963); see also, Herbert Feigl, *The 'Mental' and the 'Physical': The Essay and a Postscript* (Minneapolis: University of Minnesota Press, 1967).

[7] Margolis, *op. cit.*, Chapter 1. See also, Joseph Margolis, *Art and Philosophy* (Atlantic Highlands: Humanities Press, 1980), for a discussion of the ontology of cultural entities.

[8] Cf. Feigl, *loc. cit.*; also, Mario Bunge, 'Emergence and the Mind', *Neuroscience*, XI (1977), and Bunge, 'Levels and Reduction', *American Journal of Physiology*, CIII (1977).

[9] Cf. Popper and Eccles, *op. cit.*, Chapter P2. Eccles, incidentally, seems committed to ontic dualism, whereas Popper is rather more inclined toward attribute dualism.

[10] Cf. Hempel. *loc. cit.*: and Wesley C. Salmon, 'Statistical Explanation', in Wesley C. Salmon *et al.* (eds.), *Statistical Explanation and Statistical Relevance* (Pittsburgh: University of Pittsburgh Press, 1971), also, Causey, *op. cit.*, p. 173.

[11] The view is regularly pressed, for instance, by Donald Davidson (see below).

[12] Cf. Ernest Nagel, *The Structure of Science* (New York: Harcourt, Brace and World, 1961); and Donald Davidson, 'Causal Relations', *Journal of Philosophy*, LXIV (1967), and Davidson, 'Mental Events', in Lawrence Foster and J. W. Swanson (eds.), *Experience and Theory* (Amherst: University of Massachusetts Press, 1970).

[13] See Causey, *op. cit.*, Chapter 2; also Karl Popper, 'The Propensity Interpretation of Probability', *British Journal for the Philosophy of Science*, X (1960); and James H. Fetzer, 'Statistical Probabilities: Single Case Propensities vs. Long-Term Frequencies', in Werner Leinfellner and Eckehart Köhler (eds.), *Developments in the Methodology of Social Science* (Dordrecht: D. Reidel, 1974).

[14] Cf. 4, below.

[15] Cf. Nagel, *loc. cit.*

[16] Cf. Popper, *loc. cit.*; Fetzer, *loc. cit.*

[17] The most strenuous advocate of this view is J. J. C. Smart, *Philosophy and Scientific Realism* (London: Routledge and Kegan Paul, 1963). Cf. also, Karl R. Popper, 'The Aim of Science', in *Objective Knowledge* (Oxford: Clarendon, 1972).

[18] One of the ablest discussions of the complexities of interlevel causal explanation appears in Kenneth Schaffner, 'Theory Structure in the Biomedical Sciences', unpublished.

[19] Cf. 5, below.
[20] See Sigmund Freud, *The Interpretation of Dreams*, rev., trans. A. A. Brill (London: G. Allen and Unwin, 1948).
[21] See *From Max Weber: Essays in Sociology*, H. H. Gerth and C. Wright Mills (trans.), (New York: Oxford University Press, 1946).
[22] See Clifford Geertz, *The Interpretation of Cultures* (New York: Basic Books, 1973).
[23] See Erving Goffman, *The Presentation of Self in Everyday* Life (Garden City: Doubleday, 1959).
[24] See Hempel, *loc. cit*.; and William H. Dray, *Laws and Explanation in History* (London: Oxford University Press, 1957).
[25] A fuller discussion appears in *Persons and Mind*. See, also, *Philosophy of Psychology* (Englewood Cliffs: Prentice-Hall, 1984).
[26] Cf. Jürgen Habermas, *Knowledge and Human Interests*, Jeremy J. Shapiro (trans.) (Boston: Beacon Press, 1971).
[27] See Noam Chomsky, *Language and Mind*, enl. edn. (New York: Harcourt Brace Jovanovich, 1972).
[28] See Claude Lévi-Strauss, *The Elementary Structures of Kinship*, rev. edn., J. R. Bell, J. R. von Sturmer, and Rodney Needham (ed.) (trans.) (Boston: Beacon Press, 1969); and *The Raw and the Cooked*, trans. John and Doreen Weightman (New York: Harper and Row, 1969).
[29] See Jean Piaget, *Structuralism*, Chinanah Maschler (trans.) (New York: Basic Books, 1971).
[30] See B. F. Skinner, *The Behavior of Organisms* (New York: Appleton-Century-Crofts, 1938).
[31] See, further, *Persons and Minds*; also, 3, below.
[32] Franz Brentano, 'The Distinction between Mental and Physical Phenomena' (1874), in Oskar Kraus (ed.), *Psychology from an Empirical Standpoint*; English edn. ed. Linda L. McAlister (London: Routledge and Kegan Paul, 1973); and Edmund Husserl, *Logical Investigations*, 2 vols., J. N. Findlay (trans.) (New York: Humanities Press, 1970).
[33] Cf. James Cornman, 'Intentionality and Intensionality', *Philosophical Quarterly*, XII (1962).
[34] See John Haugeland, 'The Nature and Plausibility of Cognitivism', *The Behavioral and Brain Sciences*, I (1978). The leading (but opposed) models of the new cognitivism appear most clearly in Jerry A. Fodor, *The Language of Thought* (New York: Thomas Y. Crowell, 1975); and Daniel Dennett, *Brainstorms* (Montgomery, Vt.: Bradford Books, 1978). Cf. Margolis, *Persons and Minds*; also, 'The Trouble with Homunculus Theories', *Philosophy of Science*, XLVII (1980).
[35] Cf. Margolis, *Art and Philosophy*.
[36] Cf. Margolis, *Persons and Minds*, Chapter 9.
[37] These notions are developed in a quite different way by Pierre Bourdieu, particularly in his analysis of *habitus*; see his *Outline of a Theory of Practice*, Richard Nice (trans.) (Cambridge: Cambridge University Press, 1977). Affinities with both Merleau-Ponty and Wittgenstein are noticeably strong in Bourdieu.
[38] See for example Paul Ziff, *Semantic Analysis* (Ithaca: Cornell University Press, 1960); Hilary Putnam, *Philosophical Papers*, Vol. 2 (Cambridge: Cambridge University Press, 1975); Douglas L. Stalker, *Deep Grammar* (Philadelphia: Philosophical Monographs, 1976).

[39] Cf. Putnam, *loc cit.*; Joseph Margolis, 'The Meaning of a Word', *Metaphilosophy*, IX (1978).
[40] The fullest account of embodiment – and, also, of tokens-of-types – appears in *Art and Philosophy*.
[41] Cf. Maurice Merleau-Ponty, *The Structure of Behavior*, trans. Alden L. Fisher (Boston: Beacon Press, 1963).
[42] The token/type distinction is due to Charles Sanders Peirce, who construes it semiotically; cf. *Collected Papers of Charles Sanders Peirce*, Vol. 4, Charles Hartshorne and Paul Weiss (eds.) (Cambridge: Harvard University Press, 1939).
[43] Cf. Jack Glickman, 'Creativity in the Arts', in Lars Aagaard-Mogensen (ed.), *Culture and Art* (Nyborg and Atlantic Highlands: F. Løkkes Forlag and Humanities Press, 1976); Richard Wollheim, *Art and Its Objects* (New York: Harper and Row, 1968); and Nicholas Wolterstorff, 'Toward an Ontology of Art Works', *Nous*, IX (1975).
[44] Cf. Donald Davidson, 'Actions, Reasons, and Causes', *Journal of Philosophy*, LX (1963).
[45] David Wiggins, *Identity and Spatio-Temporal Continuity* (Oxford: Basil Blackwell, 1967); and Eli Hirsch, *The Persistence of Objects* (Philadelphia: Philosophical Monographs, 1977). Wiggins's book has been revised and enlarged and reissued as *Sameness and Substance* (Cambridge: Harvard University Press, 1980).
[46] Cf. Gadamer, *loc. cit.*, and *Philosophical Hermeneutics*, trans. David E. Linge (Berkeley: University of California Press, 1976); and E. D. Hirsch, Jr., *Validity in Interpretation* (New Haven: Yale University Press, 1967); also, Margolis, *Art and Philosophy*.
[47] See Hilary Putnam, 'Minds and Machines', in *Philosophical Papers*, Vol. 2. A fuller discussion of the inadequacy of (so-called cognitive) functionalism is provided in *Philosophy of Psychology*.
[48] I should like to express my warm appreciation to Eckehart Köhler for a number of extremely helpful editorial suggestions and queries.

2

THE CONCEPT OF CONSCIOUSNESS

The key to the problem of consciousness depends on the obvious fact that ascriptions of consciousness are introduced in significantly different, but conceptually related, ways with respect to human persons and creatures of nonhuman species or infants at phases of development antecedent to their functioning as fully competent persons. The pivotal fact, however, is a naive and simple one: among persons, that is, *among ourselves*, consciousness is introduced with the admission or recognition of a reporting role shared and understood by the members of a linguistic community. The very difficulty one has in stating this fact in a nonquestionbegging way testifies to its fundamental importance. On the other hand, consciousness is introduced in speaking of creatures that lack language, by way of an observer's attempt to explain satisfactorily the behavior and development they exhibit. Consciousness, therefore, is a theoretical posit of some sort with regard to creatures other than human and with regard to the infant phases of human life prior to the development of linguistic ability and the appearance of actual linguistic performance. It serves an explanatory role with regard to fully human persons as well. But it cannot be *introduced* in a merely explanatory way, because the very effort to explain and understand the sense in which *we* are conscious entails our being such. In our own case, we must introduce consciousness reflexively, for it is ineliminably entailed by our ability to report and share our thoughts, perceptions, feelings, intentions, and the like. Human persons, then, serve as the irreplaceable paradigms of what it is to be conscious. Man may not be the center of the universe, but he is the center of every effort to understand it.

To say this, of course, is not to say what we mean by consciousness. It is only to say that, in a certain plain sense, we cannot deny consciousness to ourselves. This perhaps is a way of recovering the Cartesian *cogito* without the dubious and puzzling certainty that Descartes originally assigned and without the pose of hyperbolic solipsism.[1] Nevertheless, we can consider the conceptual eliminability of consciousness at both the explanatory and reporting levels – always recognizing that explanation has a clear linkage with the reporting use of language. B. F. Skinner's attempt at a radical behaviorism[2] sought to eliminate, at the level of explaining the molar behavior of humans and other organisms, all use of mental and intentional predicates (except perhaps as a shortcut for the replacing vocabulary). But then, Skinner may be taken to have intended to provide an analysis *of* consciousness – in certain nonmental and nonintentional ways – rather than to insure its elimination. In a more formal way, W. V. Quine attempted to sketch the general replacement of the intensional features of intentional discourse by an idiom that behaves fully extensionally,[3] but there is no sense in Quine's program of any need to talk of eliminating mental phenomena – *a fortiori*, consciousness.

In any case, there is very good reason to believe that neither Skinner nor Quine succeeded in their projects. Skinner's efforts remained almost entirely a program of promises; Skinner seems not to have been entirely aware of the difficulty of his undertaking.[4] And Quine's sketches of a solution along extensional lines are clearly questionbegging – or else purely formal, that is, dependent on the analysis of epistemic powers that Quine never attempted to analyze.[5]

Wilfrid Sellars attempted to construct a more strenuous account of persons solely in terms of "roles" somehow added or assigned to selected aggregates of microtheoretical entities allegedly adequate for the explanation of all phenomena.[6] Though he does not eliminate consciousness altogether, Sellars seems to deny the reality of conscious states – treats them ascriptively (or "forensically") only. Bolder efforts to repudiate consciousness and mental states altogether have never been more than programmatic. Two of the principal lines of exploration have been sketched by Richard Rorty[7] and Paul Feyerabend.[8] But neither has come to grips with the reporting use of language[9] and with the reporting aspect of efforts at scientific explanation itself. In a word, we understand what it would be like to attempt to eliminate all reference to consciousness and mental states, but no viable program for their actual elimination has as yet surfaced. There is reason to believe that the difficulties to be overcome are not merely technical but are difficulties both of principle and of "real time" limitations.[10]

The study of consciousness, then, begins reflexively. This in itself has surprisingly powerful consequences. First of all, it can be shown that the analysis of the mental states of animals is inherently anthropomorphized, since the relevant categories formed for the *explanation* of animal behavior and existence are bound to be conceptually parasitic on just those categories that jointly bear on characterizing the reporting abilities of human beings and the reflexive explanation of their characteristic behavior. Secondly, it can be shown that, since the paradigms of conscious life are supplied by the human ability to report and understand one's thoughts and feelings, all ascriptions of consciousness and mental states are linguistically modelled, even for creatures lacking language. This, of course, is a corollary of the anthropomorphized nature of animal studies. Thirdly, no methodological skepticism (as distinct from error) is possible with regard to the correctness of our description and explanation of animal minds, simply because animals that are not linguistically competent share no reporting function among themselves and pursue no efforts at a theoretical understanding of themselves or of others. Thomas Nagel's otherwise interesting query, "What is it like to be a bat?"[11] is, therefore, off the mark in this regard. Nagel raised the question in order to throw into doubt both a reductive materialism that sought to eliminate interior states of consciousness and an anthropocentric account of consciousness that failed to admit the possibility of alien forms of life. But Nagel's question is the wrong one, for *only* human persons (on our present view of the population of the universe) could answer his question. He might have asked, "What is it like to be a dolphin?" *if* he supposed that we have as yet failed to crack the dolphin *language*; or, "What is it like to be a Martian?" which only Martians (on the assumption) could at the present time answer.

In virtue of these considerations, one sees that the theory of consciousness is bound to be unified – ranging over humans and animals – precisely because of the methodological differences between how consciousness is introduced in human and animal contexts; for those differences confirm the priority of reflexive characterizations and the analogical extension of relevant predicates to a range of creatures lacking self-consciousness. We may in general restrict self-consciousness to language-using creatures, that is, to creatures capable of reflexive reference and reflexive predication because of their mastery of language. We may permit a certain small tolerance here, say, among chimpanzees, where minimal self-reference seems quite possible[12] – perhaps even among domesticated dogs, where training seems to invoke responses incipiently like shame and guilt. In effect, we may concede that, as in the chimpanzee's apparent ability to attend to herself through her own mirror image, there is probably some biological predisposition toward self-reference – perhaps very much attenuated or at least distantly analogous even in the behavior of dogs – presupposed in the very achievement of linguistic self-reference. The tantalizing clues, for instance, offered by David Premack's experiments with the chimpanzee use of language seem too strong to justify dismissing self-reference altogether.[13] Perhaps at the level of nonhuman animals, a distinction is needed in order to contrast the full sell-reference of humans (involving rich second-order episodes) and the ability of certain animals to refer, distributively, in a minimally extensional sense to "themselves" – or even augmented bit by bit by certain intentional details.[14]

Self-consciousness appears to be conceptually linked to the admission of a reporting role (or, more generously, to behavior taken to be functionally equivalent to actual reporting). Consciousness, on the other hand, is attributable to languageless animals on the strength of some advantage of an explanatory sort, in virtue of which theories that avoid mental states are judged deficient. Nevertheless, to admit consciousness at the explanatory level is not to construe the admission in a merely instrumentalist sense. Also, such ascriptions are not so much the adoption of a particular theory regarding the explanation of animal life (though that, doubtless, is their motivation); they entail rather the adoption of a conceptual orientation in which theories of a certain sort (centered on whatever consciousness presupposes or entails) will be favored. Speaking loosely, the admission of consciousness is tantamount to subscribing to a meta-theory or model of alternative theories of a certain range, not to any particular theory. The reason for this distinction will be pursued very shortly.

But first, we must act to block several possible sources of misunderstanding. Human beings are said to speak natural languages. There is in fact no evidence of any human stock that lacks a language (and for every known language, there are bilinguals). By a natural language, we understand a determinate language that is learned in infancy through interaction with the members of a society who actually use that language in their own transactions.[15] However quarrelsome his speculations about an innate grammar may be, even Noam Chomsky does not deny that one cannot speak merely in accord with his linguistic universals; one must actually learn a culturally determinate language – with respect to which, on Chomsky's theory, our innate linguistic competence normally (though not necessarily) manifests itself in terms of linguistic performance. Now, it is quite possible that a machine be programmed to perform linguistically. One can imagine for instance very sophisticated programs being used by the telephone company to "answer" questions about subscribers' numbers and the like.[16] It is, therefore, quite possible to extend the use of terms like "speak", "inform", "give information about", "answer", "tell", and the like, in virtue of which we would be willing to say of a machine that it *spoke* though it was not capable of consciousness or conscious states. In the human case, the conceptual linkage between speech and consciousness cannot be broken, because the human being learns a natural language and uses it reportorially in just those ways that, paradigmatically, manifest consciousness. The problem with the general theory of consciousness is, precisely, that the paradigms of consciousness are provided by the paradigms of self-consciousness.

It is because (on the hypothesis) machines do not learn language "naturally", from some prelinguistic condition (though, with sustained use, they may come to "learn" to use their language more effectively), that *we* entertain the prospect of extending the application of predicates regarding the use of language independently of the application of predicates regarding the manifestation of consciousness. It remains true, nevertheless, that the language imputed to such machines *is* a language only on the recognition or interpretation of some human person. So the ascription of linguistic ability to machines is itself conceptually dependent on the admission of a natural language. Furthermore, this dependence affects every effort we make, on the strength of machine performance, to read back an analysis of the process underlying human performance. The notions of intentionality and information, for instance, are, in the relevant sense, first introduced in the human context and then applied (in different ways) to animals and machines. There is no way known in which it could coherently be claimed that such notions may be defined first for machines and then applied to humans; or, at any rate, there is no way to do so if significant messages are admitted and interpreted.[17]

We see, therefore, that if we treat persons as creatures or systems capable of using language (and of whatever abilities that ability facilitates), artificial persons need not be conscious, though they may be; but natural persons cannot but be conscious, cannot but be capable of being conscious. The least reflection shows that related distinctions will be required with respect to the concept of learning as well. "Learning" is an extremely elastic term, often applied to chess-playing machines, for instance, that are thought not to be conscious; or even to planaria, whose "behavior" appears to be subject to conditioning – of such an extraordinary sort that the putative "memory" of trained planaria can be passed on to other untrained specimens merely by ingesting the ground-up remains of their predecessors! But again, since linguistically non-competent animals are ascribed consciousness only at the level of our explanations of their particular form of life, it is entirely possible that the predicate "learning" and similar predicates may be extended to phenomena regarding which no assumptions of consciousness obtain. This, for instance, seems to be intended (with some tendentiousness and conceptual uneasiness, it should be said) by the ethologists.[18] We may also notice that there need be no incompatibility between characterizing a system as an automaton and as conscious. Normally, because of the nature of actual machines and actual persons and animals, we are inclined to believe that the one excludes the other. But it is not inconceivable (even if it should prove false) that human life may ultimately be explained on the basis of automaton theory; and it is quite conceivable that the complexity of future machines may justify the ascription of consciousness.[19] Certainly, there is a fair sense in which, as Putnam remarks, "everything is a Probabilistic Automaton under some Description".[20]

The difficulty with this thesis lies not so much with its enitrely abstract claim as with the attempt to provide a machine analysis of the actual (incarnate) processes by which cognitively relevant "inputs" and "outputs" are related in human and animal systems.[21] And that undertaking itself raises questions about the adequacy of *any* merely infra-psychological model of human cognition and intelligence. It is, for instance, entirely possible to attribute understanding to a human agent in contexts in which what is psychologically internalized and appropriately informs that agent's behavior is suitably congruent with, but still fails to capture infra-psychologically, what the interpretative consensus of an environing society may judge to be the import of that behavior. There is nothing incoherent in this: we do not normally know (the agent as well) what the agent "has in mind" in acting as he does; and what (characteristically) we do includes much that, once publicly appraised, is assigned an import that very likely exceeds in some measure the agent's real computations. Not only is this true, but human agents actually act with the anticipation that it is true. The most dramatic evidence lies in the production and appreciation of art and literature. For example, even on the most conservative hermeneutic theory of literary texts, that favoring the recovery of "the author's original intent," there is an ineliminably elastic sense of accuracy already embedded in the dictum (Schleiermacher's): "everything in a given text which requires fuller interpretation must be explained and determined exclusively from the linguistic domain *common to the author and his original public*".[22] The context in which this remark was first enunciated – and has subsequently been pressed – makes it quite clear that we cannot normally know whether what is common to one's public (at the moment of reception and later) is also common to oneself; for the sense of the first is not primarily psychological (though it requires *someone's* psychological gift), whereas the sense of the second is essentially infra-psychological.

It should be emphasized, of course, that, since its ascription among humans is reflexively based on their reportorial abilities, consciousness must be realistically construed. Human persons actually possess the attributes of speaking and thinking and the like. No ontology that admits the dual nature of ascriptions of consciousness (already remarked) could consistently deny the reality of mental states – unless some ulterior form of eliminative materialism (or surrogate) were vindicated. Realism regarding human mental states is at least conditionally favored. On the other hand, with respect to animals and machines, it is quite possible to preserve an option between *realist* and *heuristic* ascriptions of consciousness and mental states. The issue, however, is more complicated than may appear. By "heuristic", we mean either that the ascriptions are *not* realist or that, though they *are* intended in a realist sense, the attributes ascribed are only heuristically described. The relevant systems do not actually have the properties ascribed; or, in being ascribed in the realist sense, they are described by a *façon de parler*, a metaphor, an analogy, a figurative convenience. On this view, by definition, a natural person (that is, a linguistically able system that has learned a natural language as a natural language) must actually be conscious and possess determinate mental states. This is why Sellars's forensic account of the mental states of human persons is utterly untenable: to admit human language *is* to admit its psychological reality.

Nevertheless, although animals lacking language can be ascribed consciousness and mental states only in a way that is linguistically modelled, we are not thereby driven to hold that those ascriptions are merely heuristic. On the contrary, it is entirely reasonable to claim that at least the higher mammals actually possess mental states of some sort. The *form* of the ascription of mental states to animals is heuristic, in the sense, precisely, that the *ascriptions* are linguistically modelled; but the mental states themselves are realistically attributed to the higher animals – on the strength of our explanatory needs. This is simply the consequence of the way in which we have answered Nagel's question. We alone make ascriptions of consciousness to the bat (if we choose to).[23] Doing so, we depend on analogies drawn between the nonlinguistically informed behavior of bats and the linguistically informed behavior of persons, and we model our asscriptions of the one on what is literally true of the other. Ignoring for the time being complications that result from attributing consciousness to animals – particularly regarding the puzzles of intensionality – these ascriptions may be realistically affirmed though heuristically modelled. There is no contradiction in that. In fact, they are required if animal psychology be admissible at all.

It should be mentioned, in all fairness, that there have been attempts to model the mental states of animals and humans nonlinguistically. But these – notably, for instance, D. M. Armstrong's effort[24] – have been obliged to introduce notions like "concepts" and "ideas", which, it may be argued, can be shown not to have eliminated at all an essential appeal to a linguistic model. (We shall return to the issue.) In any case, these considerations are sufficient to dismiss Daniel Dennett's claim that "a particular thing is an intentional system only in relation to the strategies of someone who is trying to explain and predict its behavior".[25] First of all, Dennett construes intentionality in a purely heuristic way. But realism with respect to the mental states of languageless animals is at least eligible, if not well-nigh impossible to deny; also, the mental states of humans cannot all be construed merely in terms of (linguistically informed) explanatory efforts – that is, merely instrumentally. Secondly, Dennett confuses inten*t*ionality with inten*s*ionality, holding that the former (in Brentano's sense) "is primarily a feature of linguistic entities – idioms, contexts"; in fact, he equates the two notions.[26]

Now, it is conceivable that our very model of linguistic reporting and our explanation of the mental life of languageless animals betray us in some sense – perhaps "ideologically" – into thinking that the phenomena of either or both animal and human life are correctly analyzed in terms of that model. For example, both ways of speaking emphasize the molar unity of the agent of speech and mental states, and this may be construed as the result of an understandable bias of a certain reflexive habit. Some theorists (notably, structuralists like Claude Lévi-Strauss) actually seem to hold that the underlying structures of human existence are somehow "there" in the external world – independent of the molar bias and psychological capacities of reflexively competent human agents – in spite of the fact that those structures are themselves distinctly intentional in nature.[27] There is reason to think that such an approach is either incoherent or else an extremely attenuated (and somewhat irrelevant) form of caution regarding the finality of any explanatory scheme (including Lévi-Strauss's own) or regarding the inescapability of yielding to the tendentious historical currents of one's own time.[28] If so, it may be safely dismissed. The point of mentioning it is to suggest a certain closure regarding our question, since we should then have considered both *infra*- and *supra*-human alternatives to the central role of the mental states of human persons.

It is also possible to attempt to replace the molar agents of speech and mental states with a community of "sub-personal" homunculi, rather in the manner in which Dennett has worked.[29] But, apart from the difficulties of explaining how the apparently molar person is to be completely replaced by such a molecular community, the characterization of homunculi is quite explicitly and frankly dependent on the idiom suited to the reporting role of human persons. Certain *entities*, it is true – thoughts and pains and the like – are denied actual existence; but that is hardly equivalent to denying that the having of certain attributes really obtains at the personal level. Only if persons themselves are replaced by molecular homunculi, can the realist ascription *to* persons (of mental states and consciousness) be denied; but then, the question remains whether, with whatever adjustment may be required, appropriate ascriptions to the members of the molecular community will not, in their turn, be tantamount to the admission of consciousness and mental states. The matter needs to be considered more carefully. In any case, the structuralist alternative threatens to remain a mystery, since the very conditions on which its validity depends seem inaccessible on the theory. And Dennett's alternative depends on appraising just how successfully it can replace the intuitively familiar model of persons and other molar cognitive agents. We should, therefore, carefully consider what we claim in claiming that a system is conscious.

We have already conceded that many attributions normally reserved for conscious agents may be ascribed to machines that are not conscious. We have, for instance, conceded that this holds for predicates regarding speech, acquiring and transmitting information, responding to questions, learning, adjusting behavior to new information, and the like. It looks as if a similar extension could be justified for all the verbs of propositional attitude: to calculate, to solve, to decide, to test, to conjecture, and the rest. Possibly the single recalcitrant range of predicates in which, barring consciousness, we should not be inclined to permit the extension covers bodily sensations, mental images, "raw feels". These are notoriously slighted in Gilbert Ryle's attempt to segregate the "categories" of the mental and the physical, so as to preclude interaction.[30] They are also stressed, in Herbert Feigl's notable effort to vindicate some form of reductive materialism, as constituting the most difficult conceptual barrier to any reductive program.[31] They are said as well to be able to be simulated in a functional or informational sense by machine systems that actually lack such phenomenal features.[32] Most recently, they have been reaffirmed (without argument, it must be said), by Karl Popper and John Eccles, and taken by them to commit us to some sort of Cartesian dualism.[33]

There is a clear sense in which the *having* of pain and images is inadmissible except on the assumption of consciousness. Nevertheless, (i) the ascription of the relevant phenomenal states is paradigmatically managed by way of the reflexive reporting ability of human persons; and (ii) organisms or other systems may conceivably be (and be judged to be) conscious, in spite of lacking such phenomenal states. Bodily sensations appear inextricably linked, conceptually, to states of consciousness that are not similarly characterizable by such phenomenal predicates; furthermore, such sensations and images are not demonstrably necessary in systems said to exhibit consciousness.

These considerations confirm that ascriptions of consciousness cannot convincingly be grounded in the use of any particular mental predicate: for every such use, there is the risk of an extended or attenuated or metaphorical or functionally equivalent application that actually precludes, or ignores, consciousness; and phenomenal ascriptions (as of pain) seem to be contingent manifestations of consciousness. The only reasonable analysis recommends linking ascriptions of consciousness to the use of an entire system of related predicates under empirically restricted conditions. The system must apply initially to the reportorial performances of human persons and must be first extended to the explanation of the behavior of certain higher animals. The properties of such a system are so distinctive that invoking them must be seen as constituting a provisional challenge to the usual forms of reductive materialism. For the sake of a fair presentation, therefore, we must consider in turn: (a) the conceptual features of that system; (b) the methodological defense of invoking such a system; and (c) the empirical criteria in accord with which such a system is actually invoked.

We have neutralized the extension of the relevant predicates to nonconscious machines by noticing that that use requires an interpretation on the part of fully conscious persons. This is not simply because all ascriptions are made by conscious persons; on the contrary, they may be made by machines as well. It is rather because the extension of the relevant predicates entails a comparison between applying them to nonconscious systems and to the paradigm instances that are ourselves. Ascriptions of mental states, therefore, need not be coextensive with ascriptions of consciousness, but they are nevertheless conceptually dependent on such ascriptions. First of all, mental ascriptions are initially made of human persons capable of reporting and explaining their own mental states – who, then, serve as the paradigms of consciousness. Secondly, conscious mental states are theoretically ascribed by fully human persons to animals and human infants on the grounds of explanatory advantage – in spite of their lack of language. Thirdly, mental states, or analogues of mental states, are, by extension, ascribed to nonconscious machines – again, by human persons gauging functional similarities between men and machines. And finally, unconscious mental states are ascribable only to systems capable in principle of conscious mental states. The same system of predicates, therefore, is differentially applied to human persons, languageless animals, and prelinguistic infants. Hence, it is but a step to concede that it can be extended to nonconscious machines as well. Purely verbal practices, then, can hardly be counted on to sort those conditions that we take to be necessary (but not sufficient) for consciousness and those we take to serve as criteria for actual ascriptions of consciousness. The latter exhibit a characteristic informality. Also, it cannot be straightforwardly shown that the meaning of the relevant predicates has been altered simply by extending the range of their application to nonconscious machines: intension and extension are not linked in such a simple way.[34]

The most salient – perhaps the most inclusive – necessary condition for ascriptions of consciousness may be titled rationality. By rationality, we mean a model of the coherence that must hold among the internal mental states of a system and its behavior. Rationality is not an independent criterion of consciousness. In fact, on the foregoing argument – since consciousness is ineliminably entailed at the level of human reporting – considerations of rationality are bound to apply in the same dual sense in which consciousness is itself ascribed and in accord with the same paradigms. It may serve as a criterion of consciousness, applied distributively in an infra-species sense among humans; and it may serve as a criterion of the consciousness of other species. The point at stake is that rationality is a holistic notion – the notion of a certain coherence within a system of some complexity.[35]

Here, a considerable number of distinctions are called for. For instance, we must sort out nested schemata of rationality invoked in making particular ascriptions to human persons – that range from the most inclusive species-specific characteristics to culturally more determinate and historically more variable specifications, to even more determinate, more idiosyncratic, and more personal elements. The rationality of living creatures varies from species to species; but only in the case of humans will there appear as well infra-species schemata of rationality that vary widely and diachronically and alone provide the determinate marks of rationality within the species. Here, the primary question is not one of consciousness/utter lack of consciousness but rather of alternative forms of rationality/irrationality within the acknowledged consciousness of the species. The most generic constraints among all species concern interrelationships among such elements as wants, desires, beliefs, perceptions, intentions, and actions. Ascribing an intention normally entails ascribing a belief that what is intended does not yet obtain and that what is intended falls within the agent's power to achieve. Again, ascriptions of desires or wants normally entail ascribing a general congruence between intentions and actions – informed by intervening perceptions and beliefs – and such desires or wants. In short, the ascription of mental states and consciousness conforms with the viability of the species; the pattern of ascription is basically the same for humans and for creatures of other species, except that the mode of life of the different species varies. Thus, that the great cats are carnivores must be built into our explanatory picture of the species-specific rationality of such creatures. In the human case, our schema normally accommodates some selection of the so-called prudential interests of man, which bridge the animal and the culturally groomed personal interests of human beings.[36]

One important consequence of admitting a species-specific model of rationality is that it becomes logically impossible to ascribe a particular mental state or cognitively informed action without invoking the model in support of other suitably congruent ascriptions. Random members of a species cannot be said to perceive, for instance, unless they can also be said to want or desire, and to be able to act or to intend or to try to act in accord with their wants or desires and their perceptions and beliefs. Variations and idiosyncratic limitations may appear in the lives of particular individuals (for instance, among the insane), but the model is invoked to explain the viability of the species itself. In short, the ascription of particular mental states and particular actions is intelligible only on the assumption that they fit congruently among an entire network of related states and actions. There are no atomic perceptions or beliefs. To determine that a mental state or psychologically informed action obtains is to view the life of an organism under the model of some suitably species-specific rationality. Rationality, therefore, presupposes (and, on an explanatory theory, services) the biological viability of a species. Here, then, is the methodological motivation for introducing rationality: it is an inference to the best *model*, or scheme, of explanation for given species, not merely an inference to determinate explanations under such a model. Also, the imposition of the model entails that would-be mental episodes must be construed as relata within a system of relata rather than as entirely independent phenomena. Among humans, among the paradigms of consciousness, relatively extensional criteria of particular mental states and mental episodes are still available – for example, in interpreting physical movements and neurophysiological changes as bearing a certain psychological import (in behavior and thought); but such interpretations are themselves still governed by that model, and psychological ascriptions can always be made without any one-to-one correlation to physical movements or neurophysiological changes, that is, solely in accord with the coherence requirements of the model itself. This goes decidedly contrary to the usual requirements of reductive materialism.[37] In effect, conceding the holism of rationality *and*, therefore, of mental ascriptions in accord with that model, extensional treatments of the mental cannot but be conceptual conveniences authorized in the name of that very model – *if* reductionism fails.

Another consequence is that the irrationality of individual behavior presupposes, and falls within the range of, the minimal rationality of the species. The model requires only a characteristic congruence among belief, desire, action, and the rest. Particular episodes may well be defective; but to think or to behave irrationally *is* to think or to act – which presupposes the minimal, species-specific competence or rationality of the stock. The problem arises primarily – perhaps nearly exclusively – among human beings. Consequently, there is a strong temptation to construe rationality in the human context as conformity to some determinate ideology or doctrine. A particularly obvious illustration is afforded in a recent account, by Philip Pettit, of so-called "rational man theory". Pettit confuses the general rationality of man with adherence to the constraints of decision theory. For instance, he holds that "Every human action springs from a desire or set of desires which, in view of the agent's beliefs, it promises to satisfy" – where "satisfy" is defined in accord with the thesis that "Beliefs and desires lead to action by familiar rational principles it is the job of decision theory to spell out".[38] Others may equate rationality with utilitarian calculation.[39] A Kantian would insist that rationality entails conformity with the Categorical Imperative. But the exposure of particular ideologies as of a sort of pretended constraint on rationality as such[40] does not quite bear on the grounds for first favoring explanations in terms of the coherence of the mental. In short, there *is* no model of mental states and actions that does not provide for the characteristic coherence that holds among relevant ascriptions; but to concede that is also to concede a model of *species-specific* rationality. To perceive, for instance, is to make a discrimination of some kind suitable for guiding action with respect to interest and desire. Putatively isolated perceptions utterly unrelated to desire and action would have no place in promoting the viability of the species – and would be unrecognizable, being linked to no empirically detectable advantage. It is in our effort to understand the survival of the mode of life of given species that we first postulate mental states. That we do so confirms the essential theoretical focus of our own powers of reflexive reporting.

A third consequence of our model is that mental states and actions, and their congruity, are ascribable at the molar level of organismic life, not at any molecular level – that is, not at any level at which a sub-system of a person or sentient animal is designated – unless derivatively from molar ascriptions themselves. Persons think, feel, perceive, believe, and act – not their brains or sequences of neurophysiological processes. Here, the condition of rationality must be viewed in terms of *intentionality*, or psychologically real information. Despite the uncertainties of Franz Brentano's recovery of the concept of intentionality,[41] a number of pertinent generalizations may be offered about its use. Intentionality is a sufficient mark *either* of the mental *or* psychological or of what may be done or made by psychologically endowed agents. For instance, machines and artworks exhibit intentionality merely as the products of human work and behavior, in spite of (normally) lacking mental states themselves.[42] Secondly, wherever psychological states are cognitively construed, their intentional characterization involves propositional objects at least. Thus, to perceive (in the cognitive sense), or to believe, is to perceive, or believe, *that p*, even if nonpropositional objects may also be correctly ascribed. In accord with the reflexive paradigm of human mental life, one may report *that* one perceives, or believes, this or that; and our theory of the intelligibility of such reporting requires that what is reportable in first-person contexts be, in a public way, at least in general, ascribable in third-person contexts as well.[43]

The propositional content of human perception and belief is, in being reportable, directly conveyed by suitably selected *sentences: S* sees, and reports that he sees, *that there is a horse on the hill*. The propositional content of his perception and belief is *linguistically modelled*; but to perceive and believe is not necessarily to exercise any linguistic ability or to be disposed to. Insofar as *any* creature or system is ascribed cognitively qualified states, propositional content must be assigned or assignable to such states – even if, seeing that some person is thinking, we have no idea what he is thinking about; or, seeing that some creature is poised to move, we have no idea what purpose its action may serve. This still remains the only way the coherent processes of cognitively competent systems are articulated. Hence, *if* the higher animals are supposed to perceive or desire, or to act in a way informed by perception – or in any sense that is cognitively apt – then it is impossible to avoid ascribing to them propositionally qualified mental states. We may treat propositions as heuristic entities (in the sense already provided); but only the *modelling* of the propositional content of the mental states of animals is heuristic; the states are real enough if they are real at all. This goes entirely contrary to those theorists who wish to concede that animals are aware of pain and perception but who deny that they are capable of belief or of any form of thinking.[44] On the thesis here developed, such a view is incoherent. The very idea that a lion sees an eland in a cognitively relevant way signifies that (roughly) his seeing *that* an eland is near is a discrimination pertinently linked to his characteristic appetite and behavior (that are similarly propositionally informed). Such propositional ascriptions are functionally assigned *on the assumption that the explanation of the lion's behavior requires a model of species-specific rationality*. The propositional content assigned is an artifact of the explanatory theory. But that animals are taken to have propositionally qualified mental states is jointly a consequence of the anthropomorphized nature of making mental ascriptions of animals and of our inability to construe mental states in terms other than those appropriate to the rational ordering of the appropriate relata. On a realist theory of scientific explanation, we must suppose that languageless animals do possess mental states if they are required by the best explanation; and on the best interpretation, such states exhibit intentionality or "aboutness".

The puzzles of intensional contexts that affect our analysis of human mental states[45] do not affect discourse about the mental states of animals in the same way. The reason is elementary: animals lack language, and intensionality is a property of linguistic systems. In applying our model to languageless animals, *we* must decide what *conceptual* limits and capacities best characterize each species. Mice and pigeons are known to be capable of discriminating certain perceptual forms. The behavior of dogs supports the thesis that, functionally, a dog can discriminate the presence of its master; but, even if a dog's master is the mayor of Philadelphia, there is no reason to think it can distinguish the mayor of Philadelphia as such. Hence, certain intensional constraints *are* imposed on the cognitive capacities of animals, but only by way of specifying the *concepts* they can "work" with; languageless animals are incapable, on the hypothesis, of whatever intensional distinctions linguistic ability itself presupposes. In short, attributing actual consciousness to creatures entails ascribing propositional content to a range of mental states; but, relative to species-specific rationality, such ascriptions are constrained by imputed conceptual limitations appropriate to the species. Hence, (i) *contra* Armstrong, the very admission of concepts is tantamount to the adoption of a strategy of linguistic modelling; and (ii) like propositions themselves (with which they are inextricably bound), concepts may be construed as heuristic entities.

What we see, then, is that the entire apparatus of rationality and of propositionally qualified mental states is invoked to explain the life of organisms and similar molar systems. The congruence that must hold if ascriptions of rationality have any point holds initially among the actions and various states of mind of *persons and animals*. Paradigmatically, it holds of persons capable of reflexive reporting; on a favorable theory entertained by persons, it is extended to animals whose functional life appears suitably similar to the life that human beings can report. *Information*, from this point of view – not in the sense of communication theory[46] – is intentional in nature. It cannot be ascribed at any molecular level within an organismic or molar system, unless derivatively. Dennett summarizes the issue rather neatly: "the information or content an event [has] within the system [it has] for the system as a (biological) whole".[47]

By analogy with what we have already said, however, it is entirely possible to extend the use of the concept of rationality – or to use it figuratively or heuristically – among lower animals (conceivably even among plants, certainly among machines), to which we should deny mental states. What this amounts to is construing the relatively invariant "responsiveness" of creatures like the tick[48] as exhibiting, at least by a *façon de parler*, a form of rationality comparable with that of creatures actually capable of cognitively qualified mental states. In short, there is a conceptually ordered declension of ascriptions from the fully cognitive to the functional or teleological that utterly lacks cognitive qualification.[49] At the level of human persons, we admit mental states that are (a) conscious, (b) rationally interrelated, (c) cognitively qualified, and (d) linguistically informed. At the level of the higher animals, we admit mental states exhibiting (a)–(c), where the relevant ascriptions are anthropomorphized in the manner described. Such systems are said (as by Charles Taylor) to be fully *purposive* – and intentional. By various restrictions, we may distinguish alternative functional systems. For example, because machines are actually programmed by human persons, they may be said to exhibit, as artifacts, a characteristic form of rationality. This is simply to say that, on a human interpretation, machines may be ascribed – with or without consciousness – mental, even cognitive, states. Whether or not this must be a mere *façon de parler*, it involves (at least at the present stage of our technology) a noticeable extension in the absence of *biological* analogy. Such systems may be termed *artifactual* (if purposive), so as to emphasize their conceptual dependence on the originating purposes of human beings – which is not to deny, of course, that machines may introduce propositionally significant innovations that no human agent could achieve in real time. Alternatively put, they may be artifactual though purposive. But we also use the purposive idiom purely figuratively, as in speaking of a thermostat.

This is emphatically not true of the higher animals: only the form of ascription is borrowed, not the real states ascribed. In the case of lower animals, either (a) and (c) are coextensive – not in the sense that every instance of sentience is cognitively informed (which is not true even of humans), but in the sense that every *mode* of sentience provides occasions of cognition – or noncognitive forms of sentience may be conceded (or irritability at least) that functionally resemble the rational organization of fully conscious mental states. Such systems are said (as by Taylor) to be *teleological* rather than purposive, the richer ascriptions being a mere *façon de parler*. Teleological ascriptions resemble purposive, or psychologically informed, ascriptions because they are also holistic and because they appear to require peculiar lawlike relations linking processes and end-states that govern such processes. Whether the teleological can be replaced by nonteleological causal regularities we need not examine here.[50] But teleological explanation – as in homeostatic systems – does employ a model of information processing. Either the practice is a mere *façon de parler*, as in the case of the thermostat; or the model is heuristically invoked, as in predicting the behavior of a chess-playing machine or a plant in difficult terrain. What is peculiar at the level of DNA[51] is simply that information is sometimes ascribed as if sub-molar processes were actually autonomous.

We are now in a position to say what we mean by consciousness. *Consciousness is* (1) *the state of any paradigm system capable of using language or of influencing behavior by internal states that that system can report*; and (2) *the state of any nonparadigm system that is suitably analogous to the relevant states of paradigm systems*. This will seem outrageously simple (and therefore, doubtless, false). But it is surprisingly supple and plausible and can be shown to avoid certain serious difficulties. Dennett, for example, divides the concept of consciousness thus:

(1) A is aware$_1$ that p at time t if and only if p is the content of the input state of A's "speech center" at time t.

(2) A is aware$_2$ that p at time t if and only if p is the content of an internal event in A at time t that is effective in directing current behavior.[52]

His intention is to distinguish between reference to "privileged access" (awareness$_1$) and reference to "control" (awareness$_2$). He is persuaded that the two notions "*wrongly coalesce* in our intuitive grasp of what it is to be conscious of something".[53] But there are certain obvious difficulties. First of all, his (1) and (2) may be satisfied by systems that are not conscious – if we grant what has already been said about the extention of predicates to machines.[54] Secondly, the condition that *p* (that is, some proposition) be the content of *A*'s speech center, or effective in controlling behavior, is intelligible only on the grounds that *p* may (or some *q* to which *p* is dependently linked) be suitably assigned as the content of some actual mental state; but, on the argument offered, this cannot be defended unless *A* is a paradigm of a conscious system, or a system ascribed mental states by suitable analogy with such a paradigm. Hence, Dennett's account is either circular or questionbegging. Thirdly, the assignment of *p* as the relevant content of some internal state or internal event is intended by Dennett to apply to the parts of *A* rather than to *A* itself; but, on our argument (and Dennett's own, as cited above), the content or information assigned to an internal event within, or states of an internal part of, a system must "be understood as a function of the function within the whole system of that event [or state]".[55] In short, assignments of propositional content to molecular or sub-molar systems are intelligible only if conceptually dependent on assignments made to the molar systems of which they are themselves functional parts.

Dennett seems to agree with this. But his own reductive program requires that he analyze "a person into an organization of subsystems (organs, routines, nerves, faculties, components – even atoms) and [attempt] to explain the behavior of the whole person as the outcome of the interaction of these subsystems",[56] Hence, his own analytic program violates the principle (his own principle) that the function of a subsystem of a person is assignable only as the subfunction of the functioning of a molar person. But persons cannot be eliminated in favor of a community of molecular homunculi, for at least two reasons: (a) the ascription of propositional content to the parts has no *empirical* basis except in terms of analyzing the subfunctioning ingredients of the conscious functioning of their molar system; and (b) on the thesis, the information ascribed to the parts is *inaccessible* to the molar system. Dennett himself says: "We have no direct personal access to the *structure* of *content*ful events within us".[57] But if we cannot reduce persons by sub-personal homunculi (because homunculi are only relata within a molar system), then we cannot reduce networks of intentional homunculi to nonintentional physical systems (because we cannot treat such holistic systems extensionally).

The resilience of our definition of consciousness presupposes the distinctive emergence of the life of our paradigms. Two constraints have proved essential: first, intentional "content cannot be *described*",[58] only *assigned* on a model of rationality; and second, there is no known way to replace, for all systems, such intentional assignments by nonintentional ones.[59] But the very paradigm of intentional models is just our own conscious life, the life of human beings.

NOTES

[1] Cf. Joseph Margolis, '"I Exist",' *Mind*, LXXIII (1964); '"I Exist" Again', *Mind*, LXXX (1971).

[2] B. F. Skinner, *Science and Human Behavior* (New York: Macmillan, 1953).

[3] W. V. Quine, *Word and Object* (Cambridge: MIT Press, 1960).

[4] See Joseph Margolis, *Philosophy of Psychology* (Englewood Cliffs: Prentice-Hall, 1984), Chapter 3; also, Noam Chomsky, Review of B. F. Skinner, *Verbal Behavior*, *Language*, XXXV (1957); and Hugh Lacey, 'The Scientific Study of Linguistic Behaviour; A Perspective on the Skinner-Chomsky Controversy,' *Journal for the Theory of Social Behaviour*, IV (1974).

[5] Joseph Margolis, 'The Stubborn Opacity of Belief Contexts', *Theoria*, XLII (1977).

[6] Wilfrid Sellars, 'Philosophy and the Scientific Image of Man', in *Science, Perception, and Reality* (London: Routledge and Kegan Paul, 1963).

[7] Richard Rorty, 'In Defense of Eliminative Materialism', *Review of Metaphysics*, XXIV (1970). Rorty's views have changed; cf. *Philosophy and the Mirror of Nature* (Princeton: Princeton University Press, 1979).

[8] P. K. Feyerabend, 'Materialism and the Mind-Body Problem', *Reivew of Metaphysics*, XVII (1963). It is not clear how to reconcile Feyerabend's eliminative views with his scientific anarchism; cf. Paul Feyerabend, *Against Method* (London: NCB, 1975).

[9] Cf. James Cornman, 'On the Elimination of "Sensations" and Sensations', *Review of Metaphysics*, XXII (1968).

[10] See Joseph Margolis, *Persons and Minds* (Dordrecht: D. Reidel, 1978); also, Hilary Putnam, *Meaning and the Moral Sciences* (London: Routledge and Kegan Paul, 1978).

[11] Thomas Nagel, 'What is it like to be a bat?' *Philosophical Review*, LXXXIII (1974).

[12] See H. S. Terrace, L. Petitto, and T. G. Bever, 'Project Nim: Progress Report I', Columbia University, mimeograph, January 30, 1976; David Premack, *Intelligence in Ape and Man* (Hillsdale, N.J.: Lawrence Erlbaum, 1976); but also, Thomas A. Sebeok and Jean Umika-Sebeok (eds.), *Speaking of Apes* (New York: Plenum Press, 1980).

[13] Premack, *op. cit.*, particularly pp. 90–92.

[14] Some of the complexities of human self-reference are conveniently explored in G. E. M. Anscombe, 'The First Person', in Samuel Guttenplan (ed.), *Mind and Language* (Oxford: Clarendon, 1975); and Cora Diamond and Jenny Teichman (eds.), *Intention and Intentionality* (Ithaca: Cornell University Press, 1979), particularly the papers of Part I. Cf. also, Harry Frankfort, 'Freedom of the Will and the Concept of a Person', *Journal of Philosophy*, LXVIII (1971).

[15] Noam Chomsky, *Language and Mind*, enl. ed. (New York: Harcourt Brace Jovanovich, 1972).

[16] See Margaret Boden, *Artificial Intelligence and Natural Man* (New York: Basic Books, 1977).
[17] See C. E. Shannon and W. Weaver, *Mathematical Theory of Communication* (Urbana: University of Illinois Press, 1949); and Norbert Wiener, *The Human Use of Human Beings; Cybernetics and Society* (Boston: Houghton Mifflin, 1954).
[18] See N. Tinbergen, *The Study of Instinct* (London: Oxford University Press, 1969); and Konrad Lorenz, *Studies in Animal and Human Behavior*, 2 vols., Robert Martin (transl.) (Cambridge: Harvard University Press, 1970).
[19] See Kenneth Sayre, *Cybernetics and the Philosophy and Mind* (London: Routledge and Kegan Paul, 1976); and Hubert L. Dreyfus, *What Computers Can't Do* (New York: Harper and Row, 1972); Boden, *loc. cit.*
[20] Hilary Putnam, 'The Nature of Mental States', in *Philosophical Papers*, Vol. 2 (Cambridge: Cambridge University Press, 1975).
[21] Cf. Jerry A. Fodor, *Psychological Explanation* (New York: Random House, 1968); and Margolis, *Philosophy of Psychology*, Chapter 3.
[22] The translation from Schleiermacher's *Hermeneutik* is given by E. D. Hirsch, Jr., *Validity in Interpretation* (New Haven: Yale University Press, 1967), p. 200 (italics added). The problem of the impossibility of deciding whether *genres* are real and effective or only heuristically adduced appears to be unresolvable in favor of a recovery of the author's intent – in any literally infra-psychological sense. The issue provides a very good test case for the adequacy of applying machine models to human intelligence, under real-time conditions; cf. further, Joseph Margolis, *Art and Philosophy* (Atlantic Highlands: Humanities Press, 1980), Chapter 6.
[23] See Donald R. Griffin, *The Question of Animal Awareness* (New York: Rockefeller University Press, 1976).
[24] D. M. Armstrong, *Belief, Truth and Knowledge* (Cambridge: Cambridge University Press, 1973); see also, Peter Geach, *Mental Acts* (London: Routledge and Kegan Paul, 1957), and Zeno Vendler, *Res Cogitans* (Ithaca: Cornell University Press, 1972).
[25] Daniel C. Dennett, 'Intentional Systems', in *Brainstorms, Philosophical Essays on Mind and Psychology* (Montgomery, Vt.: Bradford Books, 1978), pp. 3–4.
[26] *Ibid.*, p. 3.
[27] See Claude Lévi-Strauss, *The Raw and the Cooked*, trans. John and Doreen Weightman (New York: Harper and Row, 1969).
[28] Cf. also, Jürgen Habermas, *Knowledge and Human Interests*, trans. Jeremy S. Shapiro (Boston: Beacon Books, 1971).
[29] Daniel Dennett, *Content and Consciousness* (London: Routledge and Kegan Paul, 1969); *Brainstorms.*
[30] Gilbert Ryle, *The Concept of Mind* (London: Hutchinson, 1949).
[31] Herbert Feigl, *The 'Mental' and the 'Physical'; the Essay and a Postscript* (Minneapolis: University of Minnesota Press, 1967).
[32] For instance Donald M. MacKay, *Information, Mechanism and Meaning* (Cambridge: MIT Press, 1969); also, Hilary Putnam, 'Minds and Machines', in Sidney Hook (ed.), *Dimensions of Mind* (New York: New York University Press, 1960).
[33] Karl R. Popper and John C. Eccles, *The Self and Its Brain* (Berlin: Springer International, 1977).
[34] Cf. Hilary Putnam, 'The Meaning of "meaning",' in *Philosophical Papers*, Vol. 2.
[35] Cf. Donald Davidson, 'Mental Events', in Lawrence Foster and J. M. Foster (eds.),

Experience and Theory (Amherst: University of Massachusetts Press, 1970); and John Kekes, *A Justification of Rationality* (Albany: State University of New York Press, 1976).

[36] Cf. for instance, Joseph Margolis, *Negativities. The Limits of Life* (Columbus: Charles Merrill, 1975).

[37] Cf. for instance Feigl, *op. cit.*

[38] Philip Pettit, 'Rational Man Theory', in Christopher Hookway and Philip Pettit (eds.), *Action and Interpretation. Studies in the Philosophy of the Social Sciences* (Cambridge: Cambridge University Press, 1978), p. 45.

[39] Cf. J. J. C. Smart and Bernard Williams, *Utilitarianism For and Against* (Cambridge: Cambridge University Press, 1973).

[40] Cf. Habermas, *op. cit.*

[41] Franz Brentano, 'The Distinction between Mental and Physical Phenomena', in *Psychology from an Empirical Standpoint* (ed.), Oskar Kraus; Eng. edition ed. Linda L. McAlister (London: Routledge and Kegan Paul, 1973); Edmud Husserl, *Logical Investigations*, 2 vols., J. N. Findlay (trans.) (New York: Humanities Press, 1970). Cf. also, James Cornman, 'Intentionality and Intensionality', *Philosophical Quarterly*, XI (1962).

[42] Cf. Margolis, *Art and Philosophy*.

[43] This of course trades on one of the principal themes of Wittgenstein's later philosophy. See Ludwig Wittgenstein, *Philosophical Investigations*, G. E. Anscombe (transl.) (New York: Macmillan, 1953); and P. F. Strawson, *Individuals* (London: Methuen, 1959); but also, P. T. Geach, *Mental Acts* (London: Routledge and Kegan Paul, 1957).

[44] See for instance Donald Davidson, 'Thought and Talk', in *Mind and Language*; Zeno Vendler, *op. cit.*; Norman Malcolm, 'Thoughtless Brutes', *Proceedings and Addresses of the American Philosophical AssocitION*, XLVI (1972–73); cf. further, Margolis, *Persons and Minds*, Chapter 9.

[45] See Quine, *op. cit.*; and Joseph Margolis, 'The Stubborn Opacity of Belief Contexts', *Theoria*, XLIII (1977), and 'Arguments with Intentional and Extensional Features,' *Southern Journal of* Philosophy, XV (1977).

[46] Cf. Sayre, *op. cit.*; but also, Fred I. Dretske, *Knowledge and The Flow of Information* (Cambridge: MIT Press, 1981).

[47] Dennett, *Content and Consciousness*, Chapter 4.

[48] Cf. Tinbergen, *op. cit.*; also, Jean-Claude Ruwet, *Introduction to Ethology*, Joyce Diamanti (transl.) (New York: International Universities Press, 1972).

[49] See Charles Taylor, *The Explanation of Behavior* (London: Routledge and Kegan Paul, 1964).

[50] Cf. D. Noble, 'Charles Taylor on Teleological Explanation', *Analysis*, XXVII (1966–67).

[51] See James D. Watson, *Molecular Biology of the Gene*, 2nd ed. (New York: Benjamin, 1970).

[52] *Content and Consciousness*, pp. 118–119; cited again in *Brainstorms*, p. 30, and defended.

[53] *Brainstorms*, p. 31.

[54] Cf. 'Toward a Cognitive Theory of Consciousness', *Brainstorms*, pp. 171–172.

[55] *Ibid.*, p. 163.

[56] *Ibid.*, p. 153.

[57] *Ibid*., p. 169. Dennett actually says: "So far as I can see, however, every cognitivist theory currently defended or envisaged, functionalist or not, is a theory of the sub-personal level. It is not at all clear to me, indeed, how a psychological theory – as distinct from a philosophical theory – could fail to be a sub-personal theory. So the functionalists' problem of capturing the person as subject of experience must arise as well for these cognitivist theories," *Brainstorms*, pp. 153–154.
[58] Dennett, *Content and Consciousness*, p. 83.
[59] Dennett declares much too sanguinely (but pertinently for his own program): "... although no neat synonymy or correlation between Intentional and non-Intentional sentences has been discovered or proposed, sense has been made of the lesser claim that certain types of physical entities are systems such that their operations are *naturally* to be described in the Intentional mode – and this, only in virtue ultimately of their physical organization. The force of 'naturally' here is this: although such systems are ultimately amenable to an extensional theory of their operations, their outward manifestations are such that they can be *intelligibly* described at this time, within our present conceptual scheme, only in the Intentional mode," *ibid*., p. 89.

3

ANIMAL AND HUMAN MINDS

In speaking of the consciousness of animals, affection for one's own pets is bound to be received as a mark of dangerously reduced rigor. I am tempted to offer the sanguine anecdote of my dog's symbolic behavior – how he nuzzles me in a characteristic way, at a certain hour, in order to get me to take him out; how he touches my hand with his nose, starts for the stairs, waits for me to put on my jacket, trots to the door, pauses for me to follow, nudges his leash expectantly, moves to nuzzle me again, waits at the door, and so on. Some will see in this a grand self-deception; others, a genuinely instructive specimen; still others, a convenient *façon de parler*. There's little point in mere insistence: the obvious puzzle concerns how to address the question of animal consciousness, animal communication, animal thought, animal intelligence. But more than this, the validity of the evidence favorable to significant claims of animal competence below the level of genuine language itself entails a distinctive view of the empirical nature of the issues. It is not in the least unreasonable to resist the separation of apparent questions of fact and questions of why it is that we regard particular questions as empirical questions of fact. But the issues regarding animal consciousness lend a peculiar relevance to the conceptual linkage between these two sorts of question.

The initial constraints on inquiry are decisive. First of all, linguistic communication is initially – on a dubious reading, essentially – restricted in an infra-species way to human beings. There is no terrestrial inter-species linguistic communication involving natural languages (that is, languages learned naturally), even if there is evidence of primate competence in learning the rudiments of natural languages.[1] Of course, given the biology of the primates, the admission that they are capable of learning human languages at once entails a very high level of animal consciousness and intelligence prior to such learning (the limits of which are obviously open to considerable dispute). But the very move to ascribe linguistic ability to the great apes, the possibility of inter-species linguistic communication (even nonhuman infra-species linguistic communication – after the fact of such learning) depends on the theoretical import, both enabling and restricting, of the first constraint adduced. Very simply put, we do not have the slightest idea of any specific biological restrictions that might bar extending language use beyond the human. When, for instance, he insists that language is species-specific to humans, Chomsky trades on nothing more than the fact (if it is a fact) that only humans show even an incipient tendency toward language. Should the dolphins prove to have a natural language-like communicative system that could be strengthened in the direction of a human language, terrestrial inter-species communication among (naturally) languaged creatures may, as John Lilly has always dreamed,[2] still obtain. Certainly, extra-terrestrial inter-species linguistic communication is not inconceivable; and, closer to home, the achievement of the human infant in mastering a language, naturally, is, in spite of the sometimes misleading complications of the Chomskyan theory of the innate features of human linguistic competence, evidence of (i) a remarkably powerful form of communication between prelinguistic animals and languaged human persons, and (ii) the learning of a language within the context of prelinguistic consciousness and intelligence. Correctly understood, that achievement signifies that the difference between animals and persons is not as such a difference between species – even if it is the case that, among terrestrial creatures, all but the human species are contingently precluded from developing into persons.[3] In any case, we are not at this point claiming that any terrestrial creature other than man actually possesses a language. That issue is a vexed one, to which we shall return only very briefly, after considering what is involved in attributing a mind to animals in spite of a lack of language.

A second constraint suggests itself. Being languaged, human beings cannot, in the terrestrial setting, fail to regard themselves, reflexively, as the very paradigms of consciousness and a developed mental life. It is not simply because they alone make inquiry about the mental capacities of languageless animals that their speculations are colored by their own orientation. The bias of the human effort to understand the mental life of animals is hardly like that of French and German efforts to characterize one another. It is rather, on the assumption of the absence of inter-species linguistic communication, that human speculation is inherently restricted to considering the competence of animals in terms of how, from the exclusively human point of view, other animals compare with the human paradigm. In the absence of language, there can be no animal reporting on a footing with human reporting. Consequently: (i) there is a fundamental asymmetry between ascriptions of consciousness and mental capacity to humans and animals; and (ii) there is, for that reason, an inherent and ineliminable anthropomorphism in animal psychology – attenuated, to be sure, by the human appreciation of the decidedly alien nature of putative animal capacities.

The principal key is this. Human beings cannot regard their own consciousness and mental abilities as a theoretically dubious matter; the process of inquiry itself, as well as the normal reporting abilities of humans, entails the confirmation of their mental standing. Animals, on the other hand, cannot be ascribed a measure of consciousness except on theoretical grounds – that is, on grounds that do not assume reporting, avowing, or any other communicative abilities of a distinctly linguistic sort. Once animal mentation and communication are vindicated, the human *reporting* of inter-species communication (as between my dog and myself) becomes conceptually eligible (however difficult to analyze). But the original theoretical claim must be a claim to the effect that, given the behavior and biology of a particular species, it is an inference to the best explanation to interpret the life of the members of that species in terms that permit a developed psychological characterization – including, relevantly, specific cognitive, expressive, and communicative abilities.

Here, precisely, the observation (made earlier on) regarding the conceptual linkage between factual questions and the question of why particular questions are construed as factual questions comes into play. For, we must suppose that it is itself a matter of empirical evidence whether we can or cannot satisfactorily explain animal behavior with or without the admission of an internal mental life. If animals are conscious and psychologically developed, then they are so independently of human inquiry; but that they are is a question that reflects not merely human curiosity but the asymmetry involved in proposing a fair answer, imposed by the absence of inter-species linguistic communication in the very presence of a privileged infra-species linguistic ability. In short, only human beings can characterize the animal mind, and how it is characterized cannot fail to be formulated in terms that reflexively suit the characterization of the human mind itself. These observations may be pressed in a more detailed way. But for the moment, we may content ourselves with emphasizing: (a) the conceptual asymmetry between characterizing the human and animal mind; (b) the inherent anthropomorphism of animal psychology; and (c) the distinctive sense in which theorizing about the animal mind combines empirical and philosophical considerations. Hence, if we ask ourselves, "What is it like to be a bat?" we must realize that not only we alone can answer, but *we* can answer only in terms of how we theorize about the way in which the internal life of bats compares with the internal life of man.[4]

There is, in fact, a fair sense in which all the interesting questions about animal consciousness and intelligence are versions of such questions as: How is it possible for animals, lacking language, to be conscious and intelligent? What justifies us in ascribing a particular level of consciousness and intelligence to a particular species? What is entailed in making psychological ascriptions to animals? and finally, What is the most economical model for making psychological ascriptions to ourselves as well as to animals? Here, a second set of preliminary constraints begins to make itself felt. One may, for example, always question whether, in fact, it is reasonable to say that your sudden motion *frightened* that hawk, or to say that Rover *recognized his master*. Such queries are normally treated as "internal" questions, in the sense that the scope and legitimacy of making psychological ascriptions to animals have already been conceded in principle; what remains in doubt, we suppose, is the accuracy, precision, and validity of particular such ascriptions. The constraining questions mentioned, on the other hand, all have to do, in one way or another, with the "external" issue of how we may justifiably enter for the first time upon using psychological predicates in characterizing animal life. Our first set of constraints, then, concerned the necessary peculiarities of any human effort to attempt to understand animals psychologically; the second (intended) set concerns, rather, the general uses of a psychological idiom, whether for human, animal, machine, or fictitious referents.

Here, the relevant considerations are quite instructive. For example, psychological predicates (*all* psychological predicates) may be validly employed without attributing real psychological states or psychological attributes to the referents thereby qualified. This may occur in at least two distinct ways: (a) such predicates may be used heuristically only, as in speaking of a thermostat's controlling or turning off the heat because of its registering changes in the temperature – that is, for the sake of convenience of explanation or prediction; (b) the scope of such predicates may be gradually extended to cover the real features of systems that, in theory, need not (and in fact do not) exhibit real psychological states or attributes, as in speaking of a computerized cash register's totaling the sum of a shopper's purchases or calculating the tax on such purchases. It is, for instance, quite helpful, lacking an adequate understanding of the detailed processes of plant life, to predict the characteristic development of the root system of a tree in terms of a purely heuristic model of the tree's "searching" for water and nutrients. On the other hand, in attributing sexual behavior to the male praying mantis, one is clearly extending the use of psychologically qualified predicates beyond (possibly) genuine psychological states, since (at the very least) the mantis is characteristically decapitated by the female before exhibiting his complex sexual responses.[5]

In fact, among such ethologists as Tinbergen – somewhat less consistently, Konrad Lorenz – it is characteristically maintained that, although "the ethologist does not want to deny the possible existence of subjective phenomena in animals, he claims that it is futile to present them as causes, since they cannot be observed by scientific methods. Hunger, like anger, fear, and so forth, is a phenomenon that can be known only by introspection".[6] Lorenz, on the other hand, obviously waffles on the issue: "If I am walking along with a tame greylag goose which suddenly stretches, extends its neck and softly utters a harsh warning-call, I may say 'now it is alarmed'. However, this subjective abbreviation only means that the goose has perceived a flight-eliciting stimulus and that – in accordance with the principles of stimulus-summation – its threshold values for other flight-eliciting stimuli have been markedly lowered.... In saying that the goose is alarmed, I am expressing the freely-admitted *belief* that subjective processes are taking place within the bird.... However, the scientific content of my observation is restricted to the statement that a goose which behaves in the manner described is much more likely to fly away than usual."[7] Lorenz, apparently, "feels" that the "ability to experience pleasure and sorrow ... can be attributed to higher animals", but he cannot see any way to justify the ascription on scientific grounds; and he is unwilling to justify it on the grounds of analogy.[8]

It is just at this point that the methodologically admissible maneuver of empirical inference to the best explanation offers a range of "objective" possibilities going well beyond the rather strictly behavioristic constraints by which these remarkably gifted ethologists have hobbled themselves – even more straitly, it may be said, by adhering to their notion of "innate behavior". In a somewhat different spirit, though convergently, B. F. Skinner has insisted that cognitive and, in general, psychological factors may be radically eliminated in behavioral explanation, since, for every putative causal explanation of behavior involving cognitive factors, there will always be a causal explanation involving such behavior and environmental factors (that utterly ignore the cognitive).[9]

These considerations suggest a second constraint: where, provisionally, an inference to the best explanation invites a psychological interpretation of the life and behavior of a given species, the conceptual distinction of psychological accounts themselves depends on the prospects of alternative sorts of physical or behavioral reductionism or elimination. The complete elimination of the psychological may be taken to be a very dim prospect indeed, given the argument already sketched regarding the reflexive standing of human psychological states: no eliminative program could possibly be persuasive without demonstrating (an apparently self-defeating exercise) how to eliminate (in the relevant ontological sense) the very human effort to defend the eliminative thesis.

The logical features of psychological explanation need still to be supplied; but notoriously, it is difficult – in fact, it seems impossible – to provide a nonpsychological replacement (of any sort) for the idiom in terms of which we identify phenomena as distinctly psychological. In any case, on the reductive – but not on the eliminative – view, the psychological remains real enough; the only question then to be answered concerns the nature of psychological characterization and explanation. The complications involved in making psychological ascriptions to animals depend therefore: (a) on the grounds for treating such ascriptions realistically, (b) on the possible (conceptual) irreducibility of psychological predicates and attributes, and (c) on the implications of the anthropomorphized sources of psychological description and explanation.

At this point, we must review the question of psychological categories; for, with a suitable account in hand, we may attempt to specify a third set of constraints – in an obvious sense, the most pertinent – operating on the characterization and explanation of animal life. In fact, even before venturing any explicit thesis, we may anticipate certain extraordinarily powerful limits on the explanation of psychological phenomena, whether of the human or animal sort – *if*, that is, (a) the psychological is real, (b) *sui generis*, (c) not reducible in physical terms, and (d) paradigmatically ascribed to human beings. For example, it is (it will prove) impossible to identify particular psychological states by *any* neurophysiological or brain criteria; the import or content of psychological states is, then (will prove to be), only *assignable* to particular neurophysiological processes, on independent grounds – that is, on whatever grounds, relative to our inference to the best explanation, we first introduce psychological ascriptions. In the human setting, obviously, the most detailed and refined psychological ascriptions will depend on verbal behavior and introspective reporting; in the animal setting, psychological ascriptions will be made of nonverbal behavior, under the control of an inference putatively enhancing explanation. Only the grossest correlations between particular mental activities and brain activity could, then, possibly be established. Also, notorious errors invariably obtrude in attempting such correlations: for instance, that of confusing (conflating) the ability to speak and linguistic ability – an error rampant in the literature on split-brain phenomena and the lateralization of mental functions.[10] For example, in one of the most recent overviews of the neurological literature concerned with the origins of human language, such statements as the following may be readily found: that, in spite of admitting that "language is a good deal more than speech," nevertheless "Two generalizations can safely be made with respect to the anatomical distribution of language: (1) the special machinery for language is primarily neocortical, and (2) it is localized, for the most part, in the left cerebral hemisphere"; also, "But while it might have been concluded, just a few years ago, that the right hemisphere was completely devoid of linguistic ability, it is now apparent that this view requires some modification".[11] The trouble is that *no one* has been able thus far to specify the essential features of linguistic behavior, processes, activity (as distinct from speech), in virtue of which *any* distinctive human behavior bearing on cognitive discrimination could be sorted as not linguistically informed. Without such a theory, it is, plainly, quite impossible to associate particular intellectual processes (sorted as or as not linguistically informed) with lateralized neurophysiological processes, even though differentially lateralized brain activity is well confirmed. In fact, it may be argued that the proposed opposition is quite impossible to maintain – for conceptual reasons. We shall return to the issue shortly.

In this connection, even more startling difficulties may be mentioned – all versions of another constraint, that although human language obviously has a biological foundation, we are at the present time quite unable to fix the biological and nonbiological elements of human language, *a fortiori*, the specifically biological limitations of animals. For instance, it is not known whether human language (or speech) is an exclusively neocortical function (to the extent that it may be characterized neurophysiologically) or depends on important subcortical processes: the issue affects the possible continuity or discontinuity of human speech and primate vocalization; characteristically, discussants tend to confuse this possible continuity with the continuity of linguistic and animal communication.[12] Secondly, hemispheric lateralization appears among the primates; pertinent differences in human and animal primate brains are largely concerned with comparative size; and the actual organization of the brain relative to language (in fact, relative to all conscious and cognitive processes) is not really understood at the present time or critically linked to the comparative size of the human brain and the brains of other primates.[13] Thirdly, because human language is so dramatically distinguished in terms of our ability to speak, theories of the biological (that is, innate) sources of linguistic ability are characteristically grounded in studies of the auditory discrimination of phonemes. The emphasis here, as in Chomskyan linguistics, is upon the species-specific nature of human language and human cognition, evidence of linguistic regularities at the phonological and syntactic levels – possibly even at the semantic – that are allegedly universal for natural languages, and strict biological (species-specific) limitations on linguistic variation.[14] Nevertheless, there is increasing evidence that decidedly nonlanguage-using animals – chinchillas, for instance – have been trained to differentiate quite determinate and specialized phonetic features of speech, show an ability to generalize over novel instances, and demarcate such speech stimuli (which, characteristically, do not rest on clearly demarcated acoustical segments) within the same range as humans.[15] There is, at the present time, absolutely no evidence that the demarcation of "phonetic boundaries" relevant for speech discrimination is peculiarly language-dependent or presupposes the ability to learn natural languages either naturally or otherwise. There is apparently no evidence, as Chomsky himself seems to concede, that there are any determinate linguistic universals that do not exhibit important exceptions among natural languages. And there are serious conceptual difficulties in holding that the range of relevant phonological and syntactic regularities specifically requisite for language is either minimally or characteristically programmed genetically; the ability to discriminate the requisite phonemic patterns both appears at a prelinguistic level among nonlanguage-using animals and normally requires considerable social amplification and adjustment beyond any putative programming;[16] even an extremely relaxed word order appears tolerable in a significant range of natural languages (according to Chomsky); and the semantic dimension of human language seems both hopelessly incapable of being fully platonized and unrealistically denied an essential role in determining syntactic structure.[17]

In general, the pressure of detailed biological studies of language seems to favor the conclusion: (a) that whatever is biologically innate and pertinent to language manifests itself among the prelinguistic capacities of animals; and (b) that whatever is most distinctive of the regularities of natural languages cannot be identified completely independently of the cultural experience and training of humans. The upshot is: (1) that there are (as yet) no clear biological or conceptual grounds for denying that the primates are capable of learning natural languages by a kind of forced training (that is, without growing up, from infancy, within a suitable species-specific culture); and (2) that there are, *a fortiori*, no clear biological or conceptual grounds for denying that a wide range of animals are capable of a developed mental – cognitively significant – life of some kind. The influence of the Chomskyan model has tended to obscure the dual possibility: that the emergence of language cannot be understood without reference to the biologically grounded but specifically nonbiological features of historical culture, within which humans are groomed; and that the innate biological capacities on which language itself depends are not themselves task-specifically linguistic but are shared, or continuous, with the innate biological competences of other nonlinguistic animals.

We come, therefore, to the crux of the issue: What is involved, conceptually, in attributing to animals a developed mental life in spite of their lacking language? At least since Descartes's uneasy rejection of reasoning among animals and his equally uneasy insistence that they are unfeeling automata, the question has remained a stubborn one. Descartes conceded to animals discriminations of pain, pleasure, thirst, hunger, color, taste, odor, cold, heat, anger, fear, and the like. He seems to have equivocated regarding whether animals think. For, on his most developed view, thinking involves making judgments, which he denies to animals – which encourages the thesis of automatism; but, not infrequently, he views sensation and feeling of the sorts mentioned as a kind of thinking, for he denies that they can obtain in the absence of an "intelligent substance" – hence, in the absence of some kind of intellection.[18] Now, the equivocation is instructive because it is extremely difficult to say how the difference between man and the higher apes should be characterized, if we concede that chimpanzees and gorillas at least are capable of an interesting measure of linguistic mastery; but even if we deny linguistic ability to the primates, it is not at all clear whether judgment, purposive behavior informed by perception and belief and desire, symbolic action, infra- and inter-species communication, and the like can convincingly be denied animals. Also, as far as can be determined at the present time, as already remarked, there are no critical neurophysiological criteria of the presence or absence of mental processes involving judgment or higher mental activities, whether linguistically informed or not.

In a way, the most stunning confirmation of this enormous gap in the biology of the mental is inadvertently provided by the almost scandalous reversion to dualism in the recent efforts by John Eccles and Karl Popper to discuss the mind-body problem.[19] Eccles, in particular, despite his obviously extraordinarily detailed familiarity with the entire neurophysiological literature, cannot assign any characteristic cognitive function as such to any known system of neural processing. This is not to deny that, as for instance in certain classic studies of Geschwind's, lesions in specific areas of the cortex may account for different linguistic disorders (various anomias, for example, not due to perceptual deficit of any kind).[20] But Geschwind's studies *all* have to do with significant deviations from the normal that are linked to perceivable brain disorders, once the relevant linguistic and psychological abilities are established on independent behavioral grounds. Hence, Geschwind's studies show only how anatomical disconnections between particular sensory and motor areas produce some specific disorders among the higher functions (in either humans or animals), *not* whether those functions actually obtain or where, if they do, they may be localized.[21] This limitation in the biology of the mental is rather nicely linked, as we shall see, to the inherent distinction of psychological ascriptions.

Here, the economies of speculation favoring a developed mental life among animals take two distinct forms: either (a) we may find that the behavior of at least certain animal species cannot be satisfactorily accounted for without invoking variously graded psychologically qualified capacities, which together with comparative biological considerations, fully support the inference that they are psychologically real; or (b) we may find that, although psychological predications regarding animals may be satisfactorily replaced by the use of physical and functional terms that do not employ any psychological predicates, the same reduction does not obtain for discourse about the mental states of humans. *In both cases, psychological ascriptions remain ontologically real*: in the first, in virtue of the realist import of relevant scientific explanation; in the second, in virtue of the very meaning of conceptual reduction. We may, therefore, ignore the problem of reductionism in attempting to sketch a suitable model of psychological ascriptions.[22]

What we want is a model of the psychological life of animals that is as developed as possible, short of exhibiting and presupposing linguistic ability. That would require: (a) that the requisite states be psychologically real, and (b) cognitively significant. This in itself is quite remarkable, since it signifies how little (and, of course, how much) is needed to make the case. If, therefore, it is not really possible to explain the behavior of a species without interpreting particular behavior in a cognitively pertinent way – perhaps, for example, in accounting for a dog's playing a game of chasing and retrieving a tossed stick, or a seeing eye dog's successfully guiding his master around an unusual obstacle, or the like – we should be led at once, on empirical grounds, to postulate whatever may justifiably be claimed to be conceptually entailed by cognition. There is, of course, no need to claim that all animals to which psychological states may, in some sense, be ascribed must be capable of full-fledged cognition. Thus, for example, the remarkably varied communicative patterns among the crustaceans and arachnids (including some species among both that form distinctly articulated societies) may be fairly characterized in terms of transmitting and receiving communicative "signals". They may even be said to "recognize" signals in sexual, aggressive, and similar displays, in a psychological twilight area in which cognitively significant ascriptions would be excessive, but in which some incipient sensitivity that functions rather like cognitive discrimination (probably without internal mental states) could be assigned.[23] As it happens, this very contrast is nothing more than the recovery of Descartes's alternative views of the nature of animal sensibilities.

The essential key is simply that ascriptions of cognitive ability to any species entails; (i) a certain characteristic, species-specific form of rationality; and (ii) real internal states that (apart from reductive possibilities) may be assigned propositional content. This is, also, what is required for cognitive ascriptions to humans, except that, since humans make such ascriptions reflexively, the two conditions mentioned are there linguistically informed. The point of the first condition, regarding rationality, is: (a) that mental states cannot be ascribed atomically, one by one; that individual states as of perception, volition, desire, intention, and the like can only be ascribed on relational grounds involving some suitable subset of the others; and (b) that the set of such related states must conform to some norm of coherence appropriate to a particular species. The point of the second condition, regarding propositional content, is: (c) that cognitive states are such in virtue of an agent's possessing information about some state of affairs, knowing or believing or conjecturing (or the like) that something or other is the case; and (d) that the content of such states is (in humans, normally) expressed or expressible by using declarative sentences, or (among languageless animals) functionally such that it may be modelled by such sentences. Here, we arrive quite explicitly at the meaning of an earlier claim, that animal psychology is inherently anthropomorphized. The only way to model propositions or propositional content is by means of sentences. But, since animals lack language, the determinate specification of an animal's cognitive states – as that Rover sees *that his master is at the door*, or that a hawk intends *to attack the chicken that is on the ground below* – is itself only heuristically assigned. On the hypothesis, the psychological states are real enough, but their characterization is conceptually dependent (ineliminably) on the use of a model first introduced reflexively to characterize the psychological states of linguistically competent agents. To recover these rather abstract distinctions at a stroke, one has only to consider that, in general, claiming that Rover sees his food before him signifies that he sees *that* his food is before him; and that we should not be prepared to advance that claim unless, assuming Rover hungry, of normal appetite, and the like, he acts congruently, that is, to eat the food that he sees and wants. Imputations of conceptual capacity, characteristic desires and intentions and the like are empirically based on the comparative behavior and biological endowment of the different species. This, of course, is not to say, in the behavioristic spirit, that whenever he is hungry and the like, Rover will behave congruently or that he is hungry only when he behaves congruently; or that if these or similar correlations do not obtain then mental states cannot properly be ascribed to Rover.[24]

Our model is distinctly accommodating and troublefree. For instance, we need not, in making the requisite ascriptions, assume or imply that languageless animals can think or reason in any way that requires language; correspondingly, we need not worry about the so-called intensional puzzles of human cognitive states – for example, whether, if Rover sees that his food is before him, he sees that some food is before him, or that something is before him, or that he sees either that food is before him or drink is before him, and the like. *We* provide what we regard as perspicuous characterizations of his cognitive states, and these are no more than heuristic renderings of the real functional import of aspects of his life and behavior. An animal, of course, cannot infer from one sentence to another. But, just as we justify assigning cognitive states to animals – hence, the capacity to judge that something or other is the case – we may justifiably assign inferential capacities as well. For example, if a lion sees that there is an eland nearby and begins to stalk it, then, conceding the lion's desire, intention, perceptual beliefs, and the like, we are in effect bound to construe his pattern of stalking as inferentially adjusted to shifting information about the eland's probable behavior. To see this is to disarm entirely an entire battery of misplaced doubts about whether animals can think or reason or have thoughts or reasons. They cannot, for trivial reasons, do so in the way in which humans do, because, by hypothesis, they lack language; but *if* we find that we cannot explain animal life without assuming that animals are capable of *some* cognitive states, then we cannot deny that animals think and reason and infer (exhibit strategies, future-directed intentions, emotional reactions to what they perceive, desires which they attempt to assuage), in the sense in which the functional import of their behavior cannot but be modelled on the model of thinking and reasoning paradigmatically appropriate for humans.[25] It is of course a direct consequence of accounting for animal behavior thus (and for human behavior as well) that the formal requirements of reductive materialism cannot in principal be met – once psychological ascriptions are taken as real but required, sometimes, only by constraints of coherence internal to the model of rationality.

To put the argument, now, in the simplest terms – in terms of our original anecdotal impulse – it does seem preposterous (and unhelpful from the point of view of understanding and explanation) to deny that the higher animals at least make perceptual discriminations of a cognitively relevant sort, which their finely adjusted behavior, largely dependent on contingent learning, accommodates. But if that is so, then there is no conceptually accessible way in which to deny that, in the same sense, they must be capable of a measure of intelligent adjustment among desire, belief, intention, perception, memory, learning, and action that constitutes a kind of thinking or reasoning. How these most advanced abilities must be graded and distributed through the entire range of animals depends on a kind of empirical parsimony, until finally we reach a point at which cognitive ascriptions seem inappropriate and only the barest sensitivities to stimuli may be assigned. Even at this level, perhaps, sensitivities may be graded – for example, in the span between purely instinctual behavior and behavior that involves a modicum of genuine learning.[26] But it is really the distinction between languaged humans and languageless animals that constitutes the most serious conceptual challenge to the extension of psychology within the animal domain.

Perceptual abilities are among the most difficult to deny non-languaged animals. But sensory perception, manifesting that measure of cognitive import familiar in influencing a dog's behavior – say, in the anticipatory sense of being about to be leashed for a walk – entails the assignment of paired propositional and nonpropositional objects that all perceptual cognition requires.[27] *If* a simpler model of perceptual cognition were at hand, then we could more confidently deny a kind of thinking to animals. As it is, there is no convincing alternative that eliminates propositions; and inasmuch as this is so and inasmuch as mental life is holistic, we must concede that we theorize about a dog's mind only from the distance of our own alien experience.

To admit this is both to agree with and to oppose a familiar claim advanced by Donald Davidson. For Davidson holds that there is a sense in which "thought depends on speech."[28] What we have been arguing for here, is, broadly speaking, the thesis that the concept of thought depends on the concept of speech, but that thought does not depend on speech. There is a very obvious *non sequitur* in Davidson's argument – or, at any rate, a gap that needs to be filled. For Davidson moves rather too quickly from inviting us to agree that "a speaker must himself be an interpreter of others" to inviting us to concede "the chief thesis of [his] paper . . . that a creature cannot have thoughts unless it is an interpreter of the speech of another".[29]

He cites, with some approval and disapproval, Wilfrid Sellars's observation "that thinking at the distinctively human level . . . is essentially verbal activity".[30] His objection is that thought and speech are interdependent but neither can claim "primacy".[31] But he has missed Sellars's point in an elementary way: for, for one thing, Sellars does not commit himself, here, to the impossibility of languageless thought; and for a second, Sellars is surely not saying that thinking is speaking (as in the manner of a "primitive behaviorism", which Davidson himself eschews). He is saying rather (possibly quarrelsomely) that thinking at the human level is characteristically "spontaneous thinking-out-loud".[32] In any case, Sellars's thesis is entirely compatible with the claim that the concept of thought is modelled on the concept of speech, whereas Davidson holds that there is no thinking without speaking; on the issue of whether the concept of thought depends on the concept of speech, Davidson ventures no explicit opinion. What he offers in its place is the thesis that "a creature [cannot] have a belief if it does not have the concept of belief".[33] But, for one thing, it is rather difficult to suppose that it is impossible that one dog should be afraid of another (think of Jane Goodall's film of the African wild dog), though it lacks the concept of fear; and, for another, it is not in the least certain that a dog *does not* (in some sense) have the concept of fear in perceiving another dog's fear. (Fear, on Davidson's view as on Descartes's, is a form of thought.) Nothing in the argument as Davidson presents the case precludes either possibility. It is only when he suggests that to be an interpreter of another *creature* one must subscribe to "a theory of truth that satisfies Tarski's Convention T (modified in certain ways to apply to a natural language)",[34] that his own account borders on the preposterous. In any case, the ensuing difficulties are enormous: Tarski himself despaired of applying his convention to more than a fraction of natural language;[35] not very many humans subscribe to any theory of truth ramified enough to be said to satisfy Tarski's convention; it is a convention that holds only if language is suitably extensional – which is most uncertain; and *if* animals could be said to have beliefs, desires, and fears, then *if* humans could be said to "subscribe" to the required theory, the behavior of animals would have to count as a suitable analogue of such subscribing, matching the other analogue. Obviously, Davidson's line of argument leads only to a stalemate. Furthermore, we have obviated the need to rely on such considerations by admitting frankly that the mental life of animals must be heuristically modelled on the verbal thoughts of humans.

As we have seen, the psychological model is imposed on languageless animals at the level of theoretical explanation. The model requires that particular ascriptions of mental states conform with some species-specific schema of rationality. It is only on this condition that propositional content may be imputed to particular states, or, better, that propositionally qualified states may be imputed to a creature at all – for instance, that a dog sees that there is food in his dish, wants to eat the food in his dish, intends to eat the food in his dish, acts to eat the food in his dish. But for that very reason, there is no conceivable way in which, independently of that explanatory model, the imputed propositional content of an animal's mental states may be detected neurophysiologically. The ascription of propositional content depends on a relational or holistic model of (cognitively qualified) mental states, but neurophysiological states and processes are not similarly identified or characterized.[36] For that reason, it is possibly only to *assign* informational or cognitive import to neurophysiological processes – which bears directly on the prospects of reductionism – for the model is first invoked on the grounds of the explanatory inadequacy of any simpler model. On the other hand, the line of demarcation between psychological states that lack propositional content entirely (usually thought of, among lower animals, in terms of some incipient irritability or sensation) and sub-psychological states is altogether a matter of conceptual convenience. The paradigm of the psychological – as exhibited in the reflexive linguistic reports and avowals of humans – is essentially cognitive. To speak of the psychological in the total absence of cognitive qualification is to extend the use of terms beyond their normal explanatory power, not to isolate what is most primitive and fundamental. The psychological is, in this sense, a range of emergent phenomena of a distinctly *sui generis* sort.

Finally, we may return to the intriguing question of the linguistic ability of animals, now that the more fundamental question of characterizing the psychological capacities of animals (lacking language) has been reviewed. Two themes, often confused, need to be sorted. In the first place, it is entirely possible to attribute to animals possessing a developed psychological capacity a capacity for using symbols in the absence of a linguistic capacity. It is notoriously difficult to say what a symbol is, but most discussions – which are remarkably rare and, more remarkably, quite elementary – tend to treat symbols (following C. S. Peirce) as (i) triadic, (ii) not confined to langauge, and (iii) not necessarily instituted by mere arbitrary convention. Moreover, if the ability to symbolize and to respond to symbols is treated functionally – more or less as we have been treating animal intelligence in general – then there is absolutely no antecedent reason for denying that the higher mammals at least are quite capable of symbolic behavior. The ascription of such behavior would, as before, depend on an inference to the best explanation. One of Peirce's more perspicuous definitions of "symbol" maintains that it is "a sign which is constituted a sign merely or mainly by the fact that it is used and understood as such".[37] Functionally construed, the behavior of my dog (in the opening anecdote) probably qualifies as having symbolic import: his nudging me to get me to take him out, his pausing for my compliant behavior to ensue, his nudging his leash just at the door cannot but count as a learned and obviously effective sequence of responses on both our parts. It may be thought that the apparent causal relationship among the elements in the story precludes symbolic connection; but, as a matter of fact, such "indexical" functioning is often thought to play an important role in the emergence and actual functioning of symbols themselves.[38]

Secondly, there is no doubt that the line of demarcation between genuinely symbolic behavior and merely effective behavioral association is extremely difficult to draw with certainty. And this is a distinction that bears with equal force on the division between symbolic and nonlinguistic and non-symbolic but psychologically developed abilities and between linguistic and nonlinguistic abilities. Recently, in fact, there has been a substantial revolt against the sanguine view that many of the experimental apes (Washoe, Sarah, Lana, Nim) actually had mastered the rudiments of language. But the truth is that, at the present moment, it is not so much that we see that we had been duped into thinking that chimpanzees and gorillas learned, and were able to learn, portions of human language as that we see that a great deal of the evidence was not decisive and that we are even now not entirely certain about what would be decisive.[39] There are no settled criteria of linguistic behavior for nonhuman agents – in particular, for nonhuman agents lacking a natural language. Nevertheless, one can suggest a sizeable number of considerations regarding which favorable and converging evidence is bound to increase the likelihood of being confronted with genuinely linguistic behavior – and which, in principle, is not clearly inaccessible to the primates. Perhaps a mere listing of some of these will serve our purpose: (a) the relative absence or low frequency of nonsense combinations of utterances; (b) the incidence of meaningful, context-relevant inventions among utterances; (c) the incidence of ordered sequences of utterances and behavior relevant to species-specific rationality; (d) the incidence of correct responses to metalinguistic queries isolated from otherwise normal interests and behavior; (e) evidence of abstraction from contextually restricted, or uniform circumstances, or circumstances strongly associated with characteristic interests and behavior; (f) matching of names and named items and matching of general predicates and instantiating instances, in both directions, in varying contexts and involving learned generalizations; (g) matching names and predicates and symbolic representations, in both directions, in varying contexts and involving learned generalizations;[40] (h) evidence of the relative purity of the testing of linguistic ability; (i) evidence of maximizing possibilities of choice and possibilities of error relative to purely verbal testing (in effect, increasing the likelihood of operative internal linguistic schemata); (j) spontaneous utterances describing or characterizing current situations, other than requests and in the absence of otherwise characteristic behavior or of otherwise characteristic interests and desires; (k) evidence of exchanging utterances, with the trainer, of a significantly descriptive sort; (l) evidence of spontaneous infra-species exchanges, or of significant infra-species success matching requesting utterances and appropriate responses, or of reduced infra-species success when relevant requesting utterances are temporarily inaccessible.[41] Considerations of these sorts are just the ones that most students of primate language use and are concerned to refine. They are also patently modest, manageable, concerned with what must be minimally pertinent to the mastery of language, and reasonably capable of a measure of precision with respect to which the contrast with mere problem-solving or learned association can be readily made out. There is every reason to believe, therefore, that the question of animal language is an empirical and entirely eligible question. But more than that, the very need to refine the issue along the lines suggested makes no sense unless the higher animals are already capable of a measure of intelligence in terms of which we can genuinely entertain the possibility that they could master the rudiments of language. In spite of appearances, that is by far the more profound admission.

Occasionally, as in Premack's studies, the weaker thesis is developed through pursuing the stronger. In a word, doubt about the linguistic ability of chimpanzees apparently experimenting with language tends to confirm the likelihood that chimpanzees are capable of quite advanced symbolic but nonlinguistic behavior. It is, for instance, difficult to read Premack's account of Sarah's apparently having mastered the metalinguistic function of names[42] without conceding that Sarah has some grasp of the conventional representational function of a counter with respect to an object represented – and that she can generalize regarding such functions, under conditions quite favorable to the measures (just mentioned) signifying positive evidence of language. Here, disputes are likely to arise about the minima of linguistic performance. But what is ordinarily not fully appreciated is that such disputes themselves enhance the likelihood of symbolic ability and of proto-cultural performance. Hence, even if we deny language to Sarah, we are bound to concede cognitive abilities that, in a very rich sense, constitute thinking or having thoughts; and conceding that, we are bound to go on to grasp the (now more) neutral sense in which we may claim a form of thinking even among dogs. What this ultimately suggests is that there may be some fair sense in which cognitively gifted but unlanguaged creatures may begin to share human cultures beyond being merely affected and altered by them. That a chimpanzee may acquire some rudimentary intentional capacity to *mean* something by a utterance it has been led to learn – which, admittedly, it would never exhibit in the wild – merely confirms the coherence of the notion that symbolic and proto-linguistic behavior *can* be learned by a creature whose species naturally lacks a language. But if that is so, then surely it is conceptually possible that human infants, belonging to a species that learns languages naturally, could also learn a language without innate *linguistic* competence – provided it had innate competences more powerful than those of the chimpanzee. This brings us full circle to the distinction between mere biology and biologically grounded culture.

NOTES

[1] The matter is inevitably quarrelsome and spoiled by evidence both of premature conclusions and doctrinal prejudice. Part of the trouble lies with the fact that we actually lack a clear criterion of linguistic ability among nonhuman animals, and part lies with an excessive reliance on behavioristic considerations. Cf. David Premack, *Intelligence in Ape and Man* (Hillsdale, N.J.: Lawrence Erlbaum, 1976).

[2] Cf. John Cunningham Lilly, *The Mind of the Dolphin* (New York: Doubleday, 1967).

[3] See Joseph Margolis, *Persons and Minds* (Dordrecht: D. Reidel, 1978).

[4] This corrects, in a sense, the misleading way in which Thomas Nagel poses the question, in his paper 'What is it like to be a bat?' *Philosophical Review*, LXXXIII (1974). Cf. Donald N. Griffin, *The Question of Animal Awareness* (New York: Rockefeller University Press, 1976).

[5] K. D. Roeder, 'Ethology and Neurophysiology', *Zeitschrift für Tierpsychologie*, XX (1963).

[6] Niko Tinbergen, *The Study of Instinct* (New York and Oxford: Oxford University Press, 1969 [1951]), p. 5.

[7] Konrad Lorenz, "Do animals undergo subjective experience?" (1963) in *Studies in Animal and Human Behavior*, Vol. 2, Robert Martin (transl.), (Cambridge: Harvard University Press, 1971), pp. 323–324.

[8] *Ibid.*, p. 334.

[9] B. F. Skinner, *Science and Human Behavior* (New York: Macmillan, 1953), p. 53. Cf. Hugh M. Lacey and Howard Rachlin, 'Behavior, Cognition and Theories of Choice', *Behaviorism*, VI (1978).

[10] I explore the issue, in the context of Joseph Bogen's theories, in 'Puccetti on Brains, Minds, and Persons', *Philosophy of Science*, XLII (1975). Both Bogen and Michael Gazzaniga acknowledge the distinction. Cf. J. E. Bogen, 'The Other Side of the Brain I: Dysgraphia and Dyscopia following Cerebral Commissurotomy', *Bulletin of the Los Angeles Neurological Society*, XXXIV (1969); 'The Other Side of the Brain II: An Appositional Mind', *Bulletin of the Los Angeles Neurological Society*, XXXIV (1969); and Michael Gazzaniga, *The Bisected Brain* (New York: Appleton-Century-Crofts, 1970). But neither is fully prepared to characterize the properties of the linguistic that are not confined to speech phenomena. Without doing so, however, it is utterly futile to talk of lateralizing language, to distinguish between linguistic and nonlinguistic mental operations (Bogen's special concern), or even to detect the presence of a determinately distinct physical correlate of a given mental activity.

[11] Oscar S. M. Marin, Myrna F. Schwartz, and Eleanor M. Saffran, 'Origins and Distribution of Language', in Michael S. Gazzaniga (ed.), *Handbook of Behavioral Neurobiology*, Vol. 2 (Neuropsychology) (New York and London: Plenum Press, 1979), pp. 181, 193. Although recent studies have gone some length in correcting the radical tendency to lateralize consciousness, linguistic comprehension, and other similar global functions of intelligence, there remains considerable evidence of an uncritical habit of ascribing such functions to the hemispheres themselves. Cf. for example, R. W. Sperry, E. Zaidel, and D. Zaidel, 'Self Recognition and Social Awareness in the Deconnected Minor Hemisphere', *Neuropsychologia*, XVII (1979); Jerre Levy, 'Manifestations and Implications of Shifting Hemi-Inattention in Commissurotomy Patients', in E. A. Weinstein and R. P. Friedland (eds.), *Advances in Neurology*, Vol. 18 (New York: Raven Press, 1977). The conceptual complications of making such ascriptions are, characteristically, simply ignored.

[12] Cf. Marin *et al.*, *ibid.*, p. 180. Cf. also, R. E. Myers, 'Comparative Neurology of Vocalization and Speech: Proof of a Dichotomy', *Annals of the New York Academy of Sciences*, CCLXXX (1976); and H. D. Steklis and S. R. Harnad, 'From Hand to Mouth: Some Critical Stages in the Evolution of Language', *Annals of the New York Academy of Sciences*, CCLXXX (1976).

[13] Cf. Marin *et. al.*, *op. cit.*, pp. 180–181. Cf. also, R. L. Holloway, 'Paleoneurological Evidence for Language Origins', *Annals of the New York Academy of Sciences*, CCLXXX (1976).

[14] See for example, Eric H. Lenneberg, *Biological Foundations of Language* (New York: John Wiley, 1967); Noam Chomsky and Morris Halle, *The Sound Pattern of English* (New York: Harper & Row, 1968); Noam Chomsky, *Language and Mind*, enlarged edition (New York: Harcourt Brace, Jovanovich, 1972); Edward Walker (ed.), *Explorations in the Biology of Language* (Montgomery, Vt.: Bradford Books, 1978). But see, also, the telling concessions regarding the provisional status of the theory, and the possibility that semantic and nonlinguistic experiential considerations may decisively and adversely affect any attempt to isolate (biologically) the deep syntactic structure of natural languages, in Noam Chomsky, *Language and Responsibility*, trans. John Viertel (New York: Pantheon, 1979), pp. 152–153.

[15] Cf. Marin *et al.*, *op. cit.*, pp. 186–187. Cf. also, P. K. Juhl and J. D. Miller, 'Speech Perception by the Chinchilla: Voiced-voiceless Distinction in Alveolar Plosive Consonants', *Science*, CXC (1975); and A. M. Liberman *et al.*, 'Perception of the Speech Code', *Psychological Review* , LXXIV (1967).

[16] Marin *et al., op. cit.*, pp. 188–189.
[17] See Jerry A. Fodor, *The Language of Thought* (New York: Crowell, 1975) – the most explicit attempt to develop an innatist account of human concepts; also, Jerrold J. Katz, 'Recent Issues in Semantic Theory', *Foundations of Language*, III (1967); and Margolis, *Persons and Minds*, Chapter 8. Cf. also, Joseph Margolis, *Philosophy of Psychology* (Englewood Cliffs: Prentice-Hall, 1984).
[18] The most convenient summary of this aspect of Descartes' position is provided in Zeno Vendler, *Res Cogitans* (Ithaca: Cornell University Press, 1972), Chapter 7, skewed, however, in accord with Vendler's own views.
[19] Karl R. Popper and John C. Eccles, *The Self and Its Brain; An Argument for Interactionism* (Berlin and New York: Springer International, 1977).
[20] Norman Geschwind, 'The Varieties of Naming Errors' (1967), in *Selected Papers on Language and the Brain* (Dordrecht: D. Reidel, 1974).
[21] Cf. Geschwind, 'Disconnexion Syndromes in Animals and Man', *loc. cit.*
[22] The full range of reductive possibilities – as well as their prospects of success – are canvassed in Margolis, *Persons and Minds, Philosophy of Psychology.*
[23] Cf. Peter Weygolde, 'Communication in Crustaceans and Arachnids', in Thomas A. Sebeok (ed.), *How Animals Communicate* (Bloomington: Indiana University Press, 1977); also, Jonathan Bennett, *Rationality* (London: Routledge and Kegan Paul, 1964).
[24] See for example Charles Taylor's useful criticism of the thesis offered by E. C. Tolman, in *The Explanation of Behaviour* (London: Routledge and Kegan Paul, 1964), Ch. 4; cf. E. C. Tolman, B. F. Ritchie, and D. Kalish, 'Studies in Spatial Learning, I, Orientation and the Shortcut', *Journal of Experimental Psychology*, XXVI (1946); also, Margolis, *Philosophy of Psychology*, Chapter 4.
[25] The usual difficulties alleged appear in Vendler, *loc. cit.*; Norman Malcolm, 'Thoughtless Brutes', *Proceedings and Addresses of the American Philosophical Association*, XLVI (1972–73); F. N. Sibley, 'Analysing Seeing (1)', in F. N. Sibley (ed.), *Perception: A Philosophical Symposium* (London: Methuen, 1971); Donald Davidson 'Thought and Talk', in Samuel Guttenplan (ed.), *Mind and Language* (New York: Oxford University Press, 1975). Cf. Margolis, *Persons and Minds*, Chapter 9.
[26] Cf. David Stenhouse, *The Evolution of Intelligence* (London: George Allen and Unwin, 1973).
[27] Cf. Roderick M. Chisholm, *Theory of Knowledge* (Englewood Cliffs: Prentice-Hall, 1966), Chapter 1; also, Joseph Margolis, *Knowledge and Existence* (New York: Oxford University Press, 1973), Chapter 2.
[28] Davidson, 'Thought and Talk', p. 8.
[29] *Ibid.*, p. 9.
[30] Wilfrid Sellars, 'Conceptual Change', in Glenn Pearce and Patrick Maynard (eds.), *Conceptual Change* (Dordrecht: D. Reidel, 1973), p. 82.
[31] Davidson, *op. cit.*, p. 10.
[32] *Loc. cit.*
[33] Davidson, *op. cit.*, p. 22.
[34] *Ibid.*, p. 13.
[35] Alfred Tarski, 'The Semantic Conception of Truth', *Philosophy and Phenomenological Research*, IV (1944).
[36] A related thesis is offered in Donald Davidson, 'Mental Events', in Lawrence Foster and J. M. Swanson (eds.), *Experience & Theory* (Amherst: University of Massachusetts

Press, 1970). But it is dubiously linked to a form of mind-body identity; cf. Joseph Margolis, 'Prospects for an Extensionalist Psychology of Action', *Journal for the Theory of Social Behavior*, XI (1981).

[37] *Collected Papers of Charles Sanders Peirce*, eds. Charles Hartshorne and Paul Weiss (Cambridge: Harvard University Press, 1931–1935), Vol. 2, par. 307.

[38] The ascription of symbolically significant behavior to the apes is often thought to be thrown into doubt by the lack of precision regarding the nature and conditions of such behavior. See for example, E. Sue Savage-Rumbaugh, Duane M. Rumbaugh, and Sally Boysen, 'Linguistically Mediated Tool Use and Exchange by Chimpanzees (*Pan troglodytes*)', *The Behavioral and Brain Sciences*, I (1978); together with 'Open Peer Commentary and Authors' Responses'. Nevertheless, the empirical evidence is clearly promising (625–628).

[39] See, for instance, H. S. Terrace, 'Is Problem-solving Language?' *Journal of the Experimental Analysis of Behavior*, XXXI (1979); H. S. Terrace, L. A. Petitto, R. J. Saunders, and T. G. Bevar, 'Can an Ape Create a Sentence?' *Science*, CCVI (1979); David Premack, *Intelligence in Ape and Man*; E. Sue Savage-Rumbaugh and Duane M. Rumbaugh, 'Symbolization, Language, and Chimpanzees: A Theoretical Reevaluation Based on Initial Language Acquisition Processes in Four Young *Pan Troglodytes*', *Brain and Language*, VI (1978); E. Sue Savage-Rumbaugh, Duane M. Rumbaugh, and Sally Boysen, 'Do Apes Use Language?' *American Scientist*, LXCIII (1980).

[40] Jonathan Bennett considers animal language very briefly, in *Linguistic Behaviour* (Cambridge: Cambridge University Press, 1976). But his claims are too extreme. For example, he states flatly: "A linguistic utterance is for communication in the sense that it is performed with the individual intention of getting a hearer to believe something. In infra-human displays, the 'utterance' does not manifest an individual intention, and sometimes not even an individual goal of a lower kind; what it does manifest is a species-wide analogue of intention – namely a rigid behavioural disposition whose biological *function* is to transmit information" (p. 204). Nevertheless, he himself concedes "that captive chimpanzees sometimes exhibit individual intentions to communicate seems to be beyond doubt"; he insists "only" that, in the wild, their communicative displays do not "involve any such intention" (pp. 203–204). His thesis, strictly speaking, is a *non sequitur* as far as the capacity for linguistic and nonlinguistic communicative intentions are concerned. But his account of animals in the wild – whether true or false – does explain how communication may be biologically developed without communicative intent.

There may be a deeper difficulty in Bennett's account, namely, that he has tied his theory of language too intimately to H. P. Grice's theory of speaker's meanings and speaker's intentions. Cf. H. P. Grice, 'Meaning', *Philosophical Review*, LXCI (1957); 'Utterer's Meaning, Sentence-Meaning, and Word-Meaning', *Foundations of Language*, IV (1968); and 'Utterer's Meaning and Intentions', *Philosophical Review*, LXXCIII (1969); and Joseph Margolis, 'Meaning, Speakers' Intentions, and Speech Acts', *The Review of Metaphysics*, XXVI (1973); but also, Bennett, *op. cit.*, Chapter 7. Along lines similar to those opposing Davidson's thesis, it is certainly not clear that intentions need to be linguistically restricted; or, that, in the wild, individual animals must fail to exhibit individual – even idiosyncratic – intentions; or, that animals must fail to "intend to communicate" in the wild, though capable there of having intentions (cf. pp. 203). Whatever its strengths and weaknesses in the context of speech, Grice's view of intentions has, as such, absolutely no bearing on the reasonableness of ascribing intentions – even communicative intentions – to unlanguaged animals.

[41] See David Premack and Guy Woodruff, 'Does the Chimpanzee Have a Theory of Mind?" *The Behavioral and Brain Sciences*, IV (1978).

[42] *Intelligence in Ape and Man*, Chapter 8.

4

ACTION AND CAUSALITY

"What made him insult her?" we ask. "What caused him to leave the country?" These are familiar locutions that invite – or appear to invite – causal explanations of human actions. The ease with which such questions arise and are answered suggests the ubiquity of causal explanation regarding the familiar range of human agency. They are not, in any obvious way, restricted to certain defective or deficient or metaphorically or legally extended forms of agency; they are normally entertained wherever we suppose human beings to be capable of the fullest freedom, liberty, choice, deliberate commitment in what they do. Still, there is a nagging and prolonged dispute among philosophers as to whether in principle human action – or at any rate, the actions of a so-called free agent, free actions, actions freely performed – may be explained in causal terms or must be explained in one or another contra-causal way.

We must suppose that the usual arguments pro and con are, by now, quite familiar to the disputants themselves. Certain well-known elementary errors, we may safely assume, need not be forever repeated. For example, we need not suppose that free action (if it is alleged not to be open to causal explanation) entails an actual *breach* of any causal law; *if* free acts are contra-causal, they are not contra-causal in that respect – very probably, on pain of ontic dualism.[1] Similarly, we cannot hope to maintain that free acts are contra-causal and, at the same time, identical with physical events that *are* open to causal explanation.[2] In a certain sense, it may even be a little naive to speak of settling the question whether human action is caused or not caused. For, to maintain that it is caused may very well entail that the notion of causality is extremely elastic, perhaps rather different at times from what it might be taken to signify in other contexts (say, where only the action of waves or falling stones is involved); and to maintain that it is not caused may very well entail that there is something peculiar about the notion of a human action, so that an action may be taken to be related in some special way to physical events that are caused. Both maneuvers are instructive. First of all, they suggest that there may be a conceptual linkage between what we are prepared to mean by a cause and whether or not we invoke causes in the context of human actions. Secondly, they draw attention to the curious fact that the nature of causes and actions is often left unspecified in affirming or denying that human actions are caused. This will become clearer as we proceed. At any rate, it seems quite reasonable to hold that we cannot say with confidence whether human actions are caused or not until we decide what an action is; and knowing *what* an action is, we cannot say with confidence whether things of that kind are caused until we decide what we mean in claiming that a causal relationship obtains.

The options are fairly straightforward. We may claim that human actions are nothing but physical events (of whatever complexity we please): in that case, we shall have either to concede that human actions are caused or attempt to defend some sort of causal indeterminism in the physical order of things. Or, we may claim that human actions are related in some way (other than by identity) to physical events: in that case, we shall have to explain what the relationship is, which may or may not settle the question of causality. On the other hand, we may define causality itself in terms of the propriety of applying to singular events so-called causal explanations: in that case, the question of whether actions are caused may be settled by deciding the propriety of applying the explanatory models termed causal. Or, we may define causality as a relationship discernible, at least sometimes, between actual pairs of events, a relationship obtaining in singular instances – inviting but not as such dependent on the provision of an explanation: in that case, we shall have to identify the relationship and then decide whether it may be found among actions.

These seem to be rather unimaginative alternatives. Yet, they are remarkably potent in sorting the best strategies regarding our issue. For example, when Donald Davidson holds that one and the same action may be described in four different ways – flipping the light switch, turning on the light, illuminating the room, alerting a prowler – it is difficult to understand his position unless we suppose that actions are identical with physical events and that one and the same action may be characterized correctly in both intentional and nonintentional terms.[3] This seems to be entailed, for instance, in maintaining that *turning on the light* equals *illuminating the room* and in maintaining that *flipping the light switch* equals *alerting a prowler.*[4] On the other hand, when Alvin Goldman holds that an action may be noncausally generated by another action – as when my sister's marrying conventionally "generates" the act of making me a brother-in-law to her husband – it is difficult to understand his position unless we suppose that actions are not (need not be) identical with physical events.[5] This seems to be entailed, for instance, because, on Goldman's view, one and the same physical event may be involved (somehow) in a number (an indefinite number) of actions that are different from one another.[6]

Now, it is a remarkable fact that neither Davidson nor Goldman, whose views have been much debated, ever explains why actions may be identified with physical events or what, if the identity relation be denied, actually is the relationship that holds between physical events and human actions. Since Goldman is committed to a theory of basic actions, it is not really crucial that he holds that, for non-basic actions, actions need not be causally generated; the ulterior question remains, whether basic (human) actions are caused. In that regard Goldman obviously intends to support a causal theory of action.[7] And, although others of Davidson's views confirm the likelihood of his identifying actions with physical events,[8] he offers no explicit and compelling argument either for his view about actions or for his more general view about mental and psychologically qualified phenomena and physical phenomena.

Davidson, however, has focused in an extremely useful way the distinction between singular events that enter into causal relations and the descriptions of such events and between causal relations and the explanations of causal events.[9] Ironically, the gain of those distinctions features the unresolved status of the issues here raised. Davidson agrees with C. J. Ducasse for instance,[10] that "singular causal statements entail no law and that we can know them to be true without knowing any relevant law".[11] In context, he means to hold (i) that an actual causal relation between two singular events may be detected without being able to explain the causal phenomenon itself;[12] (ii) that, sometimes, causal explanation of a singular event makes no use of causal laws;[13] (iii) that Mill was mistaken in thinking that we could be said to specify the particular cause of an event only to the extent that we specified its description adequately for explanation; (iv) that causal relations may be expressed extensionally; (v) that causal explanations behave intensionally; (vi) that Hume (and Mill) were correct in the view that causality entails that a covering law "exists" (even if we don't know what it is).[14] These claims appear to be both important and correct – with the possible exception of (iv) and (vi). But they do not speak to the question: (a) whether actions are events or physical events; (b) whether actions are caused; (c) what it is for an event to cause another; or (d) what, assuming a causal relation between actual singular events, is the nature of a causal law. These are strategic issues, a number of which are notably little investigated. Davidson himself says, quite frankly, "I have abjured the analysis of the causal relation".[15] He adds that he is, rather, exploring "the relation between causal laws and singular causal statements"; but in a sense, he attends to the *relation* between them rather than to the relata themselves. The indirect evidence is this: he assumes, without argument, that actions are caused; and he assumes, also without argument, that the covering laws of causal explanation are nomic universals.

It is in fact just the generality of causal laws that, on Davidson's view, leads us into anomaly in attempting to explain singular events in the causal way. For, *whatever* explains Caesar's death by stabbing (the singular event) cannot merely be whatever would explain the fact that a certain repeatable sequence involving death and damage to the vital organs (suitably described) is lawlike.[16] The general drift of Davidson's account – favoring Mill and Hume, for instance – is that causal laws are nomic universals. In "Causal Relations", the matter is handled sketchily. In "Mental Events", Davidson actually refers to the other in support of what he terms "the Principle of the Nomological Character of Causality": "that where there is causality, there must be a law: events related as cause and effect fall under strict deterministic laws".[17] There, his strategy concerns a version of the mind/body identity thesis – apparently strengthened by the appeal to deterministic laws.[18] But in a footnote, Davidson at once adds: "The stipulation that the laws be deterministic is stronger than required by the reasoning, and will be relaxed".[19] The principal qualification, reasonable in itself, is that "lawlikeness is a matter of degree . . . nomologicality is much like analyticity, as one might expect since both are linked to meaning". The most explicit of Davidson's remarks on laws is this: "Lawlike statements are general statements that support counterfactual and subjunctive claims, and are supported by their instances".[20]

The trouble is that these qualifications begin to bear on a central feature of Davidson's account of causality – *a fortiori*, on his causal account of actions – viz. item (vi), above (that causality entails that a covering law exists). If we "abjure" the causal relation itself and if we relax the requirements of causal laws sufficiently in the manner suggested, we run the risk of losing the apparent precision with which, on the extreme deterministic view, we mark a relation as a causal one. Arthur Danto has suggested, quite reasonably, that causality may be "semi-intensional", in the sense that, semantically, to ascribe causality is to suppose that there is some description under which the phenomena to be explained fall under an explanatory law.[21] On the other hand, of course, there is a fair argument to be mounted against the deterministic view *if* we wish to preserve free actions. Briefly put, we want to concede that human agents actually are able, at a given time, to perform either of a pair of mutually exclusive acts (at *t*, Tom is able to shave himself or refrain from shaving himself; if he shaved [refrained from shaving] himself, he could have refrained from shaving [not shaving] himself). One way of putting the point is to say that the extension of the actual and the (physically or naturally) possible cannot be the same, if human freedom is conceded.[22] Davidson's relaxed view of nomologicality makes room for this consideration – no matter what may be its most perspicuous formulation. Notice, by the way, that ascriptions of freedom, like causal laws, "support counterfactuals and subjunctive claims, and are supported by their instances".

The issue is precisely the one at stake in the well-known dispute between J. L. Austin and P. H. Nowell-Smith.[23] Nowell-Smith is anxious to treat freedom as not contra-causal – in C. A. Campbell's well-known sense (breaching causal laws); he wishes to distinguish between "indeterminism" and "self-determinism" (the latter formulated pretty much in Mill's way).[24] But he fails to be sufficiently explicit about what he means by determinism: he may mean merely that freedom is not contra-causal (without suitable defense), though he does not face the possibility that free action *may* be contra-causal in a sense not entailing a breach of causal law; or he may mean that human actions are determined, in which case he fails to resolve (as Mill notoriously does as well) the question of how freedom and determinism are to be reconciled. Austin's concern is to isolate the element of choice, of accessible alternatives, from the misleading network of deterministic relations; but he does not otherwise address himself to the question of the causes of actions.[25]

More recent writers have worried the question in other ways. Raziel Abelson, for instance, has tried to show that it is "logically impossible to capture voluntary human action in the net of causal explanation".[26] His position is not entirely easy to state. Partly, it rests on a "conceptual dualism" (*à la* Ryle).[27] But partly, it seems genuinely to fall back to the law-breaching version of contra-causal phenomena; if so, it leads to a deeper (and more indefensible) sort of dualism than Abelson intends to espouse. He remarks for instance:

> Having found, in avowals and performatives, *an irruption into the natural world* [italics added] by human agents that must be incompletely determined, if its author is to have incorrigible authority, the question arises: does this entail that human *actions* are causally undetermined as well, and if so, are they therefore inexplicable? Our answer to the first question will be affirmative and our answer to the second negative.[28]

The idea of an "irruption into the natural world" suggests an ontic dualism that is elsewhere rejected; the stronger and more promising thesis is that the *explanation* of human action cannot be causal in nature because, in relying on avowals and performatives, it invokes reasons. There remains an uncertainty, nevertheless. For Abelson holds that "an agent's reasons can only be known – indeed, can only *come into existence* [italics added] – through his avowals".[29] The first part of his remark suggests, *via* Ryle, the conceptual distinction of action-explanations; the second, once again, contra-causal breaches. Furthermore, there is the real possibility that Abelson has conflated the distinction between causal contexts and contexts of causal explanation: it does not follow, as Davidson has shown,[30] that if explanatory reasons behave intensionally, the *having of reasons* is not or cannot be a causal factor affecting behavior or action. For his own part, Davidson is quite mistaken in thinking that he has demonstrated – or, in fact, that it is even the case – that explanations by reasons is a *species* of causal explanation.[31] It is quite enough: (1) that explanation by reasons be compatible with causal explanation, though they be logically distinct from one another; and (2) that there can be causal explanations that include the having of reasons as causal factors. The difference between (1) and (2) is critical, since (1) is compatible with actions being contra-causal (without breaching causal laws), whereas (2) would entail that actions are open to causal explanation if (as seems reasonable) the having of reasons bears not only on moods, feelings, beliefs and the like but also on actions. The demonstration that (1) is false is linked, in Davidson's program, with a version of mind/body identity and (what Davidson calls) anomalous monism.[32] (2) is neutral to a variety of ontic positions regarding mind and body, actions and physical movements.

Arthur Danto provisionally entertains, in his account of "basic actions," the distinction between two kinds of causation – immanent, or agent causation and transeunt causation. The distinction was actively revived by Roderick Chisholm: Chisholm means by the distinction essentially to contrast one *event* causing another event or state of affairs (transeunt causation) and an agent's causing an event or state of affairs, including an action, basic or otherwise (immanent causation).[33] Danto, in effect, argues the eliminability of immanent causation on the grounds that transeunt causation is adequate for physiological events *and* that basic actions reduce to a series of these: "If the basic action which consists of *m* raising his arm is identical, as I have proposed it is, with a physiological series which terminates in *m*'s arm rising, then if the first exemplified immanent and the second transeunt causation, we hardly have *derived* our concept of transeunt causation from our experience of basic action".[34] Of course, if agents cannot be reduced (in any relevant sense) to a "physiological series," then *some* thesis of importance may well be collected by the doctrine of immanent causation. Danto's conclusion on the matter is this:

> ... though reactions and high-order responses [basic and non-basic actions] are not, as these are [that is, mechanisms of homeostasis and self-preservation], mechanical and spontaneous, no different concept of *causation* is required for their explanation ... it is consistent with the concept of basic actions that they should be caused.[35]

Danto means by this, for example, to insure that an agent's beliefs may serve as causes of his action. In this, his view resembles Davidson's. But then we need to know more about the relation of agents to bodies in order to decide the issue. Also, it remains unclear *what* the relationship is between the events (of transeunt causation) and the agents (of putative immanent causation). Danto worries determinism, primarily to insure that causality need not be taken to entail the hard thesis that denies the possibility of "something different occurring under given conditions than what in fact occurred".[36] Nevertheless, at the very close of his account, as at the opening, he emphasizes that human actions themselves are, effectively, unintelligible and cognitions are epiphenomena, unless they inform one another: "Except as saturated by representations, basic actions would be blind, neutral motions of the body and basic cognitions ephemeral provocations of the senses".[37] But this is to point to a common theme shared by all those who have favored, in one way or another, a contra-causal theory of action or a theory of immanent causation.[38] This, again, seems rather close to Davidson's claim (perhaps not altogether true, taken without qualification): "Action does require that what the agent does is intentional under some description, and this in turn requires, I think, that what the agent does is known to him under some description".[39] Davidson is not very explicit about unintentional, absentminded actions; his view seems to preclude *a priori* such "actions" from being genuine actions unless describable in the manner given. The point is quite important, for, *if* (as already suggested) apt speakers of a language cannot have internalized psychologically all the supposed rules of their language, then such speakers can produce linguistically perspicious remarks that they cannot have computed. Generalizing, we cannot deny that, with some frequency, agents may produce Intentionally significant behavior (normally, nonlinguistic) that are not known to them under any description and, with greater frequency, may produce actions (both linguistic and nonlinguistic) that are not known to them under descriptions that come to be interpretively favored in their own society.

Now, agent causation founders, whatever merit it has in fixing the distinction of human agency itself. Davidson has noted, crisply, the following dilemma: "either the causing by an agent of a primitive action is an event discrete from the primitive action, in which case we have problems about acts of the will or worse, or it is not a discrete event, in which case there seems no difference between saying someone caused a primitive action and saying he was the agent".[40] "The concept of *cause*", he says, "seems to play no role".[41] True enough. But that in itself goes no distance at all toward showing that human actions are caused unless it can be shown that human agency does not entail a contra-causal account (that is, an account not of the law-breaching variety). It is entirely possible to admit transeunt causation (as holding between events), to deny immanent or agent causation, *and* to deny (so far forth) that human actions are caused. All that is required is to deny (against Davidson) that "there is a fairly definite subclass of [physical] events which are actions",[42] or that human agents or persons are merely physical bodies (of whatever complexity).[43] This is not to support the thesis that human actions are not caused; only to demonstrate that nothing so far said shows that actions *are* caused. Merely admit a relationship other than identity to hold between actions and bodily movements, persons and physical bodies, and you have recovered the puzzle about causing human actions. It is important of course to realize that, in so speaking, the issue no longer depends on the following: (i) breaches of causal laws; (ii) determinism; (iii) agent causation; (iv) differences between explanation by causes and by reasons. It depends essentially on the rejection of reductionism. Nevertheless, the rejection of reductionism does not, unless by way of Cartesian dualism, entail that human actions are *not* caused, either. One cannot decide the issue without considering what factors inform human action; for, only in that way do we begin to see how elastic the concept of causality must be. Is it, for instance, strained so far that it would be conceptually preferable to replace causality with another relationship in order to account (a) for the bearing and influence of one agent's act on another's or (b) for the bearing and influence of bodily movements, processes and the like on an agent's action?

Once again, Davidson adopts the extreme position: "all primitive actions [tying one's shoelaces, pointing one's finger] are bodily movements".[44] But why is this so? How can it be shown? When Davidson considers actions more complex than putatively primitive actions, he begins by saying, "Assuming that we understand agency in the case of primitive actions, how exactly are such actions related to the rest?"[45] But in discussing the issue of primitive actions, he had concerned himself (usefully) only with showing, first, that, although certain events must take place in my brain and in my muscles in order for me to point my finger, it's not the case that "I must do something else that causes it. Doing something that causes my finger to move does not cause me to move my finger; it *is* moving my finger";[46] and secondly, that, although moving my finger involves "more than a bodily movement, namely a motion of the air", it is entirely possible to describe what I do so that it can be recognized as my action, not by ignoring the effects of what I do but by describing them in terms of my capacity to act intentionally.[47] Still, the first alternative begs the question whether the "is" in "it *is* moving my finger" is the "is" of identity; and the second presupposes but does not supply an adequate account of agency. So there is nothing for Davidson to fall back on. Hence, his otherwise interesting conclusion – "that our primitive actions, the ones we do not do by doing something else, mere movements of the body – these are all the actions there are. We never do more than move our bodies: the rest is up to nature"[48] – leads us nowhere. It may well be that primitive actions are the only actions there are. But that shows nothing about the identity of primitive actions and bodily movements or about the identity of persons and bodies.

The literature about action is so filled with quarrels of the sort canvassed that it is extremely difficult to appreciate the central problem about viewing human actions in causal terms. The problem, to put it very briefly, is at once ontological and concerned with the unity of science: the first because it bears on the tenability of various forms of mind/body reduction; the second, because it bears on what may be entailed in agreeing to treat human actions in causal terms. In this sense, the question is this: assuming that persons and their actions cannot be reduced to bodies and their movements and conceding what is normally admitted to be involved in human action, *can* we treat actions as caused or as causing actions on the part of other human agents? We must understand the question, now, as not entailing any quarrelsome doctrines regarding breaches of causal laws, extreme forms of determinism that preclude choice and the like, agent causation, and differences between reasons and causes and between explanations by reasons and by causes.

The most direct negative answer informed by the methodological puzzles posed by the social sciences seems to be advanced by writers distinctly influenced by Marx's *Grundrisse*[49] or focused on concerns similar to Marx's. There are, doubtless, a great many other currents contributing, with varying success, to the negative position. A recent, unpublished paper by Carol

Gould provides, however, a quite efficient entry into this very large literature mingling analytic and Continental currents, both within and without philosophy.[50] It should be noted, of course, that there is an increasingly complex range of phenomena lying between purely physical phenomena successfully explained at the present time by reference to the basic laws of physics and the phenomena of human societies; and these may well force adjustments in the concept of causal relations said to obtain at any of the intervening levels of discourse.

Broadly speaking, the most fundamental level of nature would, ideally, concern a set of microtheoretical entities subject to explicit nomic universals, such that all the phenomena usually canvassed in such sciences as physics, chemistry, and astronomy at least would be explicable in terms of aggregates of such entities. The trouble is that, at best, the would-be laws of this domain seem to be incapable of being formulated in any way other than statistically; there may in fact be an ontological constraint upon overtaking this peculiarity. Furthermore, as already remarked, to admit statistical laws is, in effect, to admit the possibility that (contrary to the usual view of those who make the concession in the context of the physical sciences) such "laws" may not represent (may not be able in principle to represent) any relevant invariant propensity in the domain examined – simply because there may not be any such real propensity to be found. In fact, in the context of evolutionary change, it is quite reasonable to conjecture that probabilistic biological regularities, if construed as lawlike, are bound not to exhibit any propensity toward invariance.[51] Also, as Karl Popper has argued, it may well be that the theory of fixed nomic universals waiting to be discovered in the nature of things entails an untenable form of essentialism, that, therefore, "Laws of nature are [to be] conceived . . . as (conjectural) descriptions of the structural properties of nature – of our world itself".[52] But apart from such considerations, the explanations of science may, for our purposes, be ordered roughly in the following progressive way: emerging in the biological domain, informationally (or functionally or, even, teleologically) ordered phenomena below the level of sentience; sentient but noncognitive phenomena; cognitively sentient (or purposive) phenomena; linguistically (or culturally) informed cognitive phenomena.[53] It's entirely possible that some of the elastic requirements on the concept of causality arise already below the level of human existence (e.g., as in speaking of the actions of intelligent and communicating animals).[54] In reviewing Gould's arguments, therefore, we are attempting an extreme economy regarding our central issue.

Gould's basic claim is that "while the category of causality is applicable to the relation of subject to object, it is inapplicable to the relation of subject to subject .... [That] relation of interaction, which is the primary subject matter of the social sciences ... is ... reciprocity". She means of course to count human agents as subjects and macroscopic physical objects (stones, coins, bodies and the like) as objects. Some of her supporting claims, however, are dubious or not decisive. For example, she speaks of "the incompatibility of causal or deterministic explanations with the very concept of an agent, or of an action", emphasizing in agreement with "post-Wittgensteinian action theorists," the role of choice. But choice, or human freedom in the sense of choice, is incompatible only with that form of determinism that takes the actual and the possible to be coextensive; or, more narrowly, it is incompatible with coercion. Certainly, the idea that human action involves choice raises questions about the form causality must take *if* we admit that free actions are subject to causal influence. But that is not (yet) the point of the objection. Again, she claims that "causal accounts fail to grasp yet another feature of action, namely that it is intentional ... directed towards the realization of the conscious purposes of an agent, although these ... need not be reflected upon or deliberated on in all cases." But Davidson's attempted model of the causal explanation of action succeeds, for one,[55] in reconciling intentions with causality. Also, the formulation could easily be construed as applying to animals incapable of human freedom: if so, the interesting question arises whether it is impossible to provide a causal explanation of the intelligent (but restricted) behavior of animals (say, of a lion's stalking an eland).

It may well be that adjustments of these sorts still do not touch on the fundamental objection. What *is* clear, however, is that, in admitting consciousness and intentional states, one sees that it must become increasingly difficult – eventually, quite impossible at the human level – to characterize causal relations as purely extensional or as extensional in the same sense as obtains among purely physical phenomena. If so, then it simply is not true, as Davidson maintains, that "although the criterion of agency is, in the semantic sense, intensional, the expression of agency is itself purely extensional. The relation that holds between a person and an event when the event is an action performed by the person holds regardless of how the terms are described; and we can without confusion speak of the class of events that are actions, which we cannot do with intentional actions".[56] For one thing, it is not clear (as Davidson assumes) that the individuation of actions conforms with the individuation of events: legal conventions and the like are against it; also, opposing theories like Goldman's show the viability (and even the use) of going against it.[57] Secondly, the problem about the extensionality of agency has to do not merely with the substitution of coextensive terms for the *actions* performed but has to do also with the substitution of putatively coextensive terms for *events* themselves putatively *identical with* the actions in question. For example, if Tom punches the policeman on the corner, then, if the policeman on the corner equals my nephew, it is not unreasonable to hold (extensionally) that Tom punches my nephew. But if Tom punches the policeman on the corner and if what he does (in performing an action) has as one consequence alerting some thieves in the vicinity, then there is no extensionally compelling basis on which to decide that *what he did, his action*, was identical with the *event* of alerting the thieves. Here, we see the force of resisting the identity relation between actions and events; correspondingly, we see the sense in which actions cannot be treated in what Davidson calls a "purely extensional way". These matters cannot be decided *a priori*. There is, then, a problem about whether, on intensional grounds, what is identified as an action on one description is or is not the same as what is identified as an action on another; and, if actions are not a subset of (antecedently specifiable) events, or at least if the criteria of identity of (physical) events (that are not actions) cannot serve to determine the numerical identity of actions (now, neutrally said to be events) under alternative descriptions,[58] there is a problem as to how to treat the causal explanation of actions as falling under covering laws. Laws may be intensionally qualified, but the assumption usually supporting the formulation of causal laws is just the one Davidson favors, namely, that causality behaves extensionally.

These considerations strengthen the point of Gould's objection, but they need not be decisive. The reason is twofold: first, because, rather along Davidson's lines, singular causal connections may be detected and explained without using causal laws – items (i)–(ii) of an earlier tally (p. 66); second, because the logical properties of, and the logical constraints operating on, laws are by no means clear; and third, because it is entirely conceivable that the regularities under which causal explanations are attempted may not need to be lawlike at all – they may for instance be merely stable social institutions changing, like other phenomena, under causal influence.[59]

Gould provides a third objection: "human actions and social phenomena in general cannot be understood or explained apart from the meanings which they have for agents". Here, she favors the views of Peter Winch and the hermeneutists.[60] Her claim is that "inasmuch as causal explanations take the relations among the entities which they study to be external, they cannot take into account the understanding which agents have of the meaning of their own actions or of the actions of others". It is true that theorists like Davidson have claimed that causal relations are purely extensional (which is at least part of what Gould means by "external"), but we have already seen both that causal relations involving intentions, reasons, meanings cannot be "purely extensional" *and* that causality itself may be sufficiently elastic to accommodate the fact. What *is* important is, precisely, that the admission of causality at this level of discourse entails, *on the condition that physicalism or any comparable form of reductionism is untenable,*[61] that covering laws cannot (here) be of the nature of nomic universals; also, that the regularities under which human actions may be explained need not be of the nature of laws at all – may for instance be of the nature of "covering institutions" subject to causal change. The crux is this: *if* singular causal relations may be detected without being able to explain the phenomena themselves, and if, sometimes, causal explanations of singular events make no use of causal laws – items (i) and (ii) in a previous tally – then it needs to be argued (it may well be a myth) that the admission of a causal connection entails that (known or unknown) a covering law "exists" (vi).

From this point of view, Gould's criticism may be read as permitting a certain conceptual choice: replace causality with another relation at the level of freedom, intentionality, and meaning; or liberate the notion of causality from those constraints that operate at the level of purely physical phenomena – in particular, that of being "purely extensional". Either maneuver affects the unity of science; the second more explicitly than the first, the theory of laws and the theory of causal explanation as well. Still, Gould's account of internal relations is open to quarrel. In the *causal* relation involved, say, in an agent's making a chair, "both the agent and the object", she says, "are changed in the relation"; thus the relation is an internal relation. But on that view, a firecracker's exploding may involve an internal relation as well; if so, then *all* "external" relations – causal relations *par excellence* – become internal relations. She adds that the causal relation (in producing a chair) is "an asymmetrical internal relation". Here, she apparently means that, first, the object (the chair) is produced in a way that depends on the intended use assigned, and the instruments and materials selected, by the agent himself; whereas, second, these objects constitute "objective conditions" (not efficacious causes) qualifying "the agent's activity and . . . the agent's self-change". So the relation is asymmetrical because the agent causes changes in objective materials, but those materials serve only as "conditions" for the agent's "*self*-change". That the relation is internal depends on the (as we now see) inadequate feature mentioned. But there is nothing yet provided which unconditionally requires the rejection of a causal model. Indeed, Gould really offers no satisfactory clue as to the *efficacy* of an agent's self-change. The best that we can get – and that much is extremely important – is that, should we invoke causality in the case of an agent's making a chair, we must do so in a way that preserves choice, intentions, the meaning and significance of one's work. *That* insures that causality cannot be "purely extensional", but it does not show that causality is incompatible with the explanation of human action.

The problem becomes one regarding the unity of science. *If* the unity of science is construed in terms of preserving: (a) the model of laws applied at the level of basic physics, (b) the explanatory model applied at that level, and (c) physicalism or any comparable reductionism,[62] then Gould is quite right in proposing to replace causality with another relationship – at the level at which human purposes and human freedom operate. But the price is a failure to understand the continuity of explanation at all levels of discourse and the evolution of human existence itself. If, on the other hand, the unity of science is construed, on the assumption of the inadequacy of physicalism or of any comparable reductionism, in terms of preserving the emergent distinction of phenomena at all levels of discourse, then there is no reason to disallow adjustments, as needed, in the concepts of causality, causal laws, explanation, even nonlawlike regularities suitable for causal explanation. It is an irony that Davidson's reasonable model of causal explanation involving human action[63] should have been so closely linked with reductionism and (at least a strong disposition toward) the exclusive adequacy of nomic universals. For, together with his insistence on the "purely extensional" nature of causal relations, these themes invite the reaction of an army of critics opposed to extreme determinism and committed to accommodating the implications of human freedom. Thus, Gould is inclined (implicitly) to agree with the general outlines of Davidson's account of causality, except that its limitations require its rejection where human action is concerned. Symptomatically, she holds that "even where the laws concerning human actions are understood as statistical regularities, this view interprets this as a limitation imposed by the complexity of the variables or the imperfect state of our knowledge" – hence, is deterministic at its core. But that interpretation is actually not favored by all commentators[64] and is incompatible with the implications of admitting intensional considerations regarding the individuation of actions.

The essential feature involved in extending the causal model to human action is that we may distinguish the independence of *particular* intentional mental states and the actions they inform and the interdependence of the *concepts* of such states and of such actions. The point is missed, for instance, by A. I. Melden[65] and by a good many others who have adopted a contra-causal theory of human action (in the so-called post-Wittgensteinian manner). It is obviously possible to intend to perform act A and fail, performing act B instead; and it is possible to intend to perform act A and not perform any act at all. I may intend to shoot the bear in my path and shoot a nearby rabbit instead; or I may intend to shoot the bear I am tracking and never act on that intention. The error is duly noted and corrected by Davidson. But, for his own part, Davidson fails to see that the intensional qualification of actions leads, on the best explanation, to these conclusions: (i) that actions cannot be identified with physical events or physical movements; (ii) that the individuation of actions cannot be managed in the same way as the individuation of physical events or physical movements; (iii) that causal relations involving actions cannot be "purely extensional." The admission of these findings would have led directly to: (iv) the untenability of physicalism or any comparable reductionism; and (v) the eligibility of causal explanations under covering regularities much weaker than are claimed for covering laws – whether nomic universals or statistical laws.

It looks very much as if Gould's claim about internal relations and the sense in which either an agent acts on an object or agents act on one another is said to constitute an internal relation suffer from the same mistake. One has only to admit that human purposes and human action are discriminable only within a rich context of cultural life: the "internality" required simply marks the resistance to reductionism (or at least premature reduction); otherwise, causal relations remain, as before, "external", which is to say contingently related – which is not of course the same thing as to say that causal relations are "purely extensional".

Gould has proposed to substitute for causality the concept of *reciprocity* among human agents, derived in a way considerably adjusted, under the influence of the *Grundrisse*, from Kant and Hegel. Reciprocity is to be both a principle of explanation and a "norm . . . a condition for the full realization of freedom". The notion is a useful one, though its normative function may here be disregarded; for if it obtains at all, it obtains even at levels of economic development (in the Marxist sense) at which human beings are still not completely free. The general features of reciprocity (as an explanatory principle) include the following: "reciprocal relations hold between independently existing agents" – hence, preclude reflexivity; reciprocal interaction is "constituted by the shared understanding and free agreement of the agents who enter into it" – hence, entails certain rich intentional relations, involvement in common projects and a common culture for instance; and in a reciprocal relation, "each agent acts with respect to the other on the basis of a shared understanding" – hence, provides conditions on which agents may increasingly respect the developing freedom of each other (the normative element).[66] To take a modest example: You offer me a cup of tea; and I, sharing the tea-institution, understand and accept your offer. The marvel of cases of this sort is that they bring home to us the extraordinary complexity of explaining how what is going on goes on. It shows at once the difficulty of attempting to explain human actions in terms of the explanatory models and the extensional reading of causal relations that obtain at the level of basic physics. But it fails to show – what it is supposed to show – namely, that causal explanation is incompatible with the explanation of human actions. It shows instead (what we have been laboring to make explicit) that, at the level of human action, causality is not purely extensional and that the causal regularities in terms of which human actions may be explained need not be nomic regularities. Whether human actions *can* be explained, at least at times, under covering laws depends very much on what we take to be an action and how informal we take a law to be. Here, ironically once again, we need only remind ourselves that the generally accepted features of laws (generality, support of counterfactuals and subjunctives, and support from conforming instances) are entirely compatible with the notion of a covering institution and human freedom; also, that (*contra* Davidson) there may well be a use to admitting as actions phenomena that are not, on any reasonable view, open to an intentional description accessible to the agent. If so, then Gould, too, has not been able to show that *only* reciprocity (as she construes it, contra-causally) can be invoked to explain human actions.

The picture, in short, is a good deal more complicated than those on either side of the issue have been prepared to admit. Causality may be taken to apply to the explanation of human action, but not reductively and not without an elastic reading of causal relations; and free actions require us to admit the elements of reciprocity (as sketched, above), but not in a way that precludes causality or requires that all particular actions be free actions.

NOTES

1 Cf. C. A. Campbell, 'Is 'Free Will' a Pseudo Problem?' *Mind*, LX (1951).

2 Cf. A. I. Melden, *Free Action* (London: Routledge and Kegan Paul, 1961).

3 Donald Davidson, 'Actions, Reasons and Causes', *Journal of Philosophy*, LX (1963).

4 Cf. G. E. M. Anscombe, *Intention* (Oxford: Basil Blackwell, 1957).

5 Alvin I. Goldman, *A Theory of Action* (Englewood Cliffs: Prentice-Hall, 1970).

[6] Cf. Jaegwon Kim, 'Causes and Counterfactuals', *Journal of Philosophy* LXX (1973). This goes against Donald Davidson's views about individuating actions (a subset of events); cf. Davidson, 'The Individuation of Events', in Nicholas Rescher (ed.), *Essays in Honor of Carl Hempel* (Dordrecht: D. Reidel, 1970).
[7] *A Theory of Action*, pp. 63–64. Nonbasic actions, therefore, are "generated" by basic acts that are caused (by the agent's wants and beliefs).
[8] See for instance 'Mental Events', in Lawrence Foster and J. W. Swanson (eds.), *Experience and Theory* (Amherst: University of Massachusetts Press, 1970); and Joseph Margolis, *Persons and Minds* (Dordrecht: D. Reidel, 1978), Chapter 11.
[9] These issues are developed chiefly in 'Causal Relations', *Journal of Philosophy*, LXIV (1967).
[10] Davidson cites the following of Ducasse's work: 'Critique of Hume's Conception of Causality', *Journal of Philosophy*, LXIII (1966); *Causation and the Types of Necessity* (Seattle: University of Washington Press, 1924); *Nature, Mind, and Death* (LaSalle, Ill.: Open Court, 1951), Pt. II.
[11] 'Causal Relations', 702.
[12] Contrast Arthur C. Danto, for instance: *Analytical Philosophy of Action* (Cambridge: Cambridge University Press, 1973), p. 82.
[13] 'Causal Relations', 700.
[14] 'Causal Relations', 701.
[15] 'Causal Relations', 699.
[16] Thus Davidson's interest in J. L. Mackie's account, 'Causes and Conditions', *American Philosophical Quarterly*, II (1965).
[17] 'Mental Events', p. 80f.
[18] Cf. *Persons and Minds*, Chapter 11.
[19] 'Mental Events', p. 81n.
[20] 'Mental Events', p. 92.
[21] *Analytical Philosophy of Action*, p. 98.
[22] Cf. Joseph Margolis, *Psychotherapy and Morality* (New York: Random House, 1966), Chapter 4, particularly pp. 97–98.
[23] J. L. Austin, 'Ifs and Cans', reprinted in *Philosophical Papers*, J. O. Urmson and G. J. Warnock (eds.) (Oxford: Clarendon, 1961); P. H. Nowell-Smith, *Ethics* (Harmondsworth: Penguin Books, 1954), Chapter 19.
[24] *Ethics*, p. 282.
[25] I can say, however, from a private conversation, that Austin was clearly interested in some form of contra-causal account.
[26] *Persons* (New York: St. Martin's Press, 1977), p. xii.
[27] Gilbert Ryle, *The Concept of Mind* (London: Hutchinson, 1949).
[28] *Persons*, p. 28.
[29] *Persons*, p. 28.
[30] Cf. 'Actions, Reasons and Causes'.
[31] Cf. Joseph Margolis, 'Puzzles Regarding Explanation by Reasons and Explanation by Causes', *Journal of Philosophy*, LXVII (1970).
[32] 'Mental Events', pp. 87–88.
[33] Cf. Roderick M. Chisholm, 'Freedom and Action', in Keith Lehrer (ed.), *Freedom and Determinism* (New York: Random House, 1966); also, Irving Thalberg, 'Do We Cause Our Own Actions?' *Analysis*, XXVII (1967).

[34] *Analytical Philosophy of Action*, p. 72. Danto also speaks of "the chaos of the distinction", p. 206. Cf. also, Donald Davidson, 'Agency', in Robert Binkley, Richard Bronaugh, and Ansonio Marras (eds.), *Agent, Action, and Reason* (Toronto: University of Toronto Press, 1971).
[35] *Analytical Philosophy of Action*, p. 115.
[36] *Analytical Philosophy of Action*, p. 184. Cf. Anthony Kenny, *Will, Freedom and Power* (Oxford: Basil Blackwell, 1975).
[37] *Analytical Philosophy of Action*, p. 195.
[38] Davidson's list – formulated in terms of denying that "rationalization is a species of ordinary causal explanation", being more suited to supporting a contra-causal theory of some sort – includes the following: G. E. M. Anscombe, Stuart Hampshire, H. L. A. Hart and A. M. Honoré, William Dray, Anthony Kenny, A. E. Melden ("and most of the books in the series edited by R. F. Holland, *Studies in Philosophical Psychology*"), 'Actions, Reasons and Causes', p. 685n.
[39] 'Agency', p. 12.
[40] 'Agency', p. 14.
[41] 'Agency', p. 15.
[42] 'Agency', p. 4.
[43] The latter denial is defended at length in *Persons and Minds*, where the relationship of embodiment is contrasted with those of identity and composition.
[44] 'Agency', p. 11.
[45] 'Agency', p. 18.
[46] 'Agency', p. 11.
[47] 'Agency', p. 14.
[48] 'Agency', p. 23.
[49] Karl Marx, *Grundrisse*, Martin Nicolaus (transl.) (New York: Vintage, 1973).
[50] Carol C. Gould, 'Beyond Causality in the Social Sciences: Reciprocity as a Model of Non-exploitative Social Relations', unpublished, first presented at The Boston Colloquium for the Philosophy of Science, Boston University, March 15, 1977. See also, Carol C. Gould, *Marx's Social Ontology* (Cambridge: MIT Press, 1978).
[51] Cf. Wesley C. Salmon, 'Statistical Explanation', in Wesley C. Salmon *et al.*, *Statistical Explanation and Statistical Relevance* (Pittsburgh: University of Pittsburgh Press, 1971); also, Carl G. Hempel, 'Aspects of Science Explanation', in *Aspects of Scientific Explanation and Other Essays in the Philosophy of Science* (New York: Free Press, 1965). Ernst Mayr reports a hundred evolutionary "laws" listed by Bernhard Rensch, which he says "refer to adaptive trends effected by natural selection. Most of them have occasional or frequent exceptions and are only 'rules,' not universal laws," *The Growth of Biological Thought* (Cambridge: Harvard University Press, 1982), pp. 37–38.
[52] Karl R. Popper, 'The Aims of Science,' in *Objective Knowledge* (Oxford: Clarendon, 1972), p. 196.
[53] See *Persons and Minds*; also, Joseph Margolis, 'Reconciling Freud's *Scientific Project* and Psychoanalysis', in *Morals, Science and Sociality* (*The Foundations of Ethics and Its Relationship to Science*, Vol. III), H. Tristram Engelhardt, Jr. and Daniel Callahan (eds.) (Hastings-on-Hudson: Institute of Society, Ethics and the Life Sciences, 1978).
[54] This is certainly one way to construe Herbert Feigl's candid admission of the problems of reductionism; cf. *The 'Mental' and the 'Physical': The Essay and a Postscript* (Minneapolis: University of Minnesota Press, 1958, 1967).

[55] 'Actions, Reasons and Causes'.
[56] 'Agency', p. 8.
[57] Cf. however, Joseph Margolis, review of Goldman's *A Theory of Action*, *Metaphilosophy*, V (1974).
[58] Cf. 'The Individuation of Events'.
[59] Cf. 'Reconciling Freud's *Scientific Project* with Psychoanalysis'.
[60] Cf. Peter Winch, *The Idea of a Social Science* (London: Routledge and Kegan Paul, 1963); Jürgen Habermas, *Knowledge and Human Interests*, Jeremy J. Shapiro (transl.) (Boston: Beacon Press, 1971); and Richard J. Bernstein, *The Restructuring of Social and Political Theory* (New York: Harcourt Brace Jovanovich, 1976).
[61] Reductionism is the key to Davidson's 'Mental Events'.
[62] Cf. *The 'Mental' and the 'Physical'*; also Otto Neurath *et al.*, *International Encyclopedia of Unified Science*, Vols. 1–2 (Chicago: University of Chicago Press, 1955).
[63] 'Actions, Reasons and Causes'.
[64] Cf. for instance, Wesley Salmon, 'Statistical Explanation'.
[65] A. I. Melden, *Free Action*, p. 52. Cf. also, Charles Taylor, *The Explanation of Behaviour* (London: Routledge and Kegan Paul, 1964).
[66] There is an intriguing and not merely superficial resemblance here to H. P. Grice's analysis of speaker's meaning; cf. 'Meaning', *Philosophical Review*, LXVI (1957).

5

PUZZLES ABOUNT THE CAUSAL EXPLANATION OF HUMAN ACTIONS

Discourse about causal relations appears to be conceptually anchored in two extremely different ways; (i) relations are said to be causal if they are explainable by appeal to causal laws; and (ii) relations are said to be causal if they are instances of agency or of some process thought to be suitably similar to agency. Merely to mention these alternatives is to collect an extraordinarily large nest of puzzles of a profound sort. We are, for example, hard pressed to provide the proper account of the nature of laws, causal laws, explanations, causal explanations; also, to provide the proper account of human, animal, and inanimate agency or causal process. But apart from such strenuous matters, it may be usefully noted: (a) that it is entirely possible that appeal to so-called natural laws need not directly bear on the causal explanation of a given phenomenon simply because the laws in question need not be causal laws, laws formulating a direct causal relationship between paired phenomena (even if an underlying causal connection be presupposed); (b) that to hold that explanatory models in science are specifically causal in nature may require appeal to more than the form of laws or to the form of explanations; (c) that causal relations may, on the agency paradigm, be said to be perceived or identified (as by a successful human agent) directly and independently of reference to any causal law or causal explanation. Thus, for instance, the Boyle-Charles' law for ideal gases, is, in Ernest Nagel's opinion, "not a causal law", primarily because it is expressed in terms of concomitant changes.[1] Nagel also observes, characterizing causal laws in general, that, among qualifying conditions to be met, "the relation [involved must be] an invariable or uniform one, in the sense that whenever the alleged cause occurs so does the alleged effect".[2] Whatever weakness or vagueness may be charged to this or similar conditions (which Nagel concedes), it is reasonably clear that not all familiar laws are causal laws; it is, also, doubtful that *any* laws have *ever* been formulated that, in the relevant sense, exhibit a strict invariance; statistical laws may not be causal laws, on the condition posed; and any more rigorous formulation along these lines of what it is to be a causal law would dampen even further the likelihood of providing convincing instances. At best, to construe statistical laws as causal laws in terms of invariance requires either that such laws are no more than approximations to actual laws (which is now regarded as a doubtful or untenable thesis) or else that the variations in the regularities of their instances fall within a certain range or approach a certain limit (which is clearly not strongly favored at least in the social and behavioral sciences – and is unlikely to be, if those sciences cannot be suitably reduced to the physical and biological sciences).

There are, also, problems with the alleged logical form of causal explanations. For instance, Carl Hempel holds that "*Causal explanation* is a special type of deductive nomological explanation; for a certain event or set of events can be said to have caused a specified 'effect' only if there are general laws connecting the former with the latter in such a way that, given a description of the antecedent events, the occurrence of the effect can be deduced with the help of the laws".[3] On Hempel's view, a deductive-nomological explanation "may be construed as an argument in which the explanandum is deduced from the explanans".[4] But this raises questions of at least two sorts. First of all, it may be that there are laws linking antecedent and subsequent events that are not causal laws: Hempel's condition is formulated only as a necessary constraint; what makes such a regularity a causal regularity is not supplied. Secondly, since explanations invoking statistical laws (that is, invoking statistical regularities not construed as tending universally toward some determinate invariance) cannot be deductive, such explanations cannot, on Hempel's view, be causal explanations, yet one supposes that there must be some. Thirdly, the mere form of the deductive-nomological model may be defective as an explanatory model, since there are formulable cases that would meet the conditions of the model without being admissible as explanations – *a fortiori*, without being causal explanations. For example, Sylvain Bromberger offers the example of a flagpole casting a shadow of a particular length, the length depending on the elevation of the sun: we explain the length of the shadow by reference to the sun's position, the height of the flagpole, and the laws governing the behavior of light; but though we may also infer the height of the flagpole from the shadow's length and the position of the sun, we would not view the inference as explanatory of the flagpole's height. Asymmetrical temporal relations, Wesley Salmon insists, "are crucial in this example".[5] Fourthly, as Salmon also points out, the argument model tends to lead us to suppose that, where determinism is abandoned, as modern physics seems to require, we still suppose that statistical explanation must make the explanandum at least highly probable. Yet, as he adds, "an explanation is not an argument that is intended to produce conviction; instead, it is an attempt to assemble the facts that are relevant to the occurrence of an event. There is no more reason to suppose that such a process will increase the weight we attach to such an occurrence than there is to suppose that it will lower it."[6] On Salmon's view, shifting to statistical explanation, we move to a suitable explanation insofar as we achieve "a relevant partition of an inhomogeneous reference class into homogeneous subclasses"; that is, roughly, "a reference class [the class of events to which the explanandum belongs, which, on some rule, yields reliable statistics] is homcgeneous if there is no way, even in principle, to effect a statistically relevant partition without already knowing which elements have the attribute in question and which do not ... each member of a homogeneous reference class is a random member".[7] Clearly, the notion of a statistical explanation requires an adjustment in the concept of causality, since invariances may no longer obtain. Following Hans Reichenbach,[8] Salmon favors an agency model – though one not requiring human agency; as he remarks: "Producing is, of course, a thoroughly causal concept; thus, the relation of interaction to orderly low entropy state becomes the paradigm of causation on Reichenbach's view, and the explanation of the ice cube in the thermos of water is an example of the most fundamental kind of causal explanation".[9]

Still, it is not unfair to note that the "production" Salmon and Reichenbach speak of involves a combination of considerations: (1) analogy between human or animal agency and the alleged "production"; and (2) stipulation in accord with certain favored theories of physical science. The reason for the quibble is straightforward. We cannot, except on the theory and the analogy favored, be said to perceive the "production" in question. But, in the case of human agency, we cannot even understand the rationality of an agent, the linkage between perception, intention, desire, belief, action and the like *without conceding that human agents produce states of affairs and changes in states of affairs and that, in behaving intentionally and with some regular success, they can perceive and understand that they and others have caused such states and changes to occur*. The concept of a human being, therefore, is, necessarily, the concept of a causally efficacious agent. Animal agency and the agency of inanimate processes are, by an increasingly attenuated (but not for that reason inappropriate) analogy construed as sources of causal efficacy. As the analogy of agency is weakened, the affirmation of causality depends on the explanatory importance of such processes – for instance, the appearance of a low entropy state. But though questions may always arise about whether this or that agent did or did not produce this or that effect, these questions make no sense unless we concede standard sorts of cases in which we normally do exert agency in a causal way and in which we normally see or know that we or others are exerting such agency. So a human being is (1) a creature capable of causal agency, and (2) capable of perceiving particular instances of causal agency.

Now, the single most important point, bearing on human agency and the causal explanation of human action, is that the perception or knowledge that *S* has exerted agency or produced an action does not require any knowledge or even any formulated belief or sketch of a belief of any *law*, deterministic or statistical, under which the causal process is to be explained. It may well be that there is a sense – conceivably, a trivial sense – in which to know that an agent produced an action intentionally is to be able to explain it: for instance, it may be that merely to redescribe a given action as an action one intended to produce *is* to explain it.[10] But, however generously we construe explanation, it is not necessary that, in treating an instance of human agency as causally efficacious, we make reference in any way at all – explicit or remote – to causal laws of any kind. Grant so much, and it is but a step to see that causal relations, at least instances of human agency or sufficiently similar animal agency, may not even presuppose that there are relevant covering laws – *a fortiori*, covering causal laws – under which they fall, or may not presuppose that they are or must be explainable under covering laws of some sort.

The possibility is admittedly heterodox. But consider that *if* causality is ascribed on the model of agency, then, assuming "A" to designate a human action, there is no apparent reason for holding that the following pair of propositions must be incompatible:

(i) A caused B;
(ii) There is, in principle, no formulable causal law under which (i) can be explained.

It may well be the case that there are causal laws (of some sort) covering *some* human actions and their effects; and it is very clear that, as the analogy between human agency and the causal agency of inanimate physical events and processes is invoked, the ascription of causal efficacy to factors of the latter sort is made to depend upon the accessibility, at least in principle, of covering causal laws. To mention one influential author again: Davidson concedes that "singular causal statements" may be known to be true without any knowledge of any relevant covering law; although *if* causality obtains, there must (he insists) be a covering law under which the singular causal connection falls.[11]

Davidson offers not the slightest defense for the position – all the more worth stressing, since he holds that causal connections behave extensionally whereas causal explanations (all explanations) behave intensionally. That is, *on* Davidson's thesis – that causal laws are linguistic formulations of some sort[12] – there appears to be a considerable lacuna in the argument that holds that the admission of (i) (above) entails the falsehood of (ii). Of course, if the distinction of a causal law is made sufficiently informal, it may be that the affirmation of a covering law may be trivialized; but it is difficult to countenance such a tolerant line, given the usual emphasis on nomic universals as the paradigms of covering laws and the care with which suspect statistical laws are admitted to be genuine laws (rather than mere epistemically defective approximations of "true" laws). In any event, on the strongest and most characteristic view, there appears to be no genuine argument of a compelling sort that shows a logical connection between (i) and the denial of (ii). J. J. C. Smart's sanguine view about eliminating all referring import from genuine nomic universals[13] pretty well suggests that the stricter the constraints on what would count as a causal law, the less obvious it is that (i) and (ii) could be connected in the manner sketched. On the other hand, the thesis that they are so connected suggests as well that a strongly reductive program – physicalism, for instance – gains plausibility precisely in that it dismisses the agency paradigm of causality and assimilates human agency to the processes of inanimate forces; for, as already remarked, the ascription of causal efficacy to inanimate events and processes is normally made to depend on physical theories that promise (competitively) to yield covering causal laws by reference to which the phenomena in question may be suitably explained. Others of course, on independent grounds – Karl Popper for instance[14] – reject the possibility of formulating valid natural laws. Popper does not deny that would-be laws must be formulated as universals; but if they were genuinely fixed and if they obtained in the manner Smart favors, we should, Popper thinks, be committed to an untenable form of essentialism. Effectively, therefore, Popper construes explanations in terms of universal laws of nature as mere phases of some inexhaustible effort to plumb the "depth" of a domain – in a sense "hardly susceptible of logical analysis".[15] He modifies essentialism "radically", therefore; though, ironically, he seems to preserve an attenuated form of the covering-law model of explanation.

There is no doubt that appeal to the phenomenon of human agency as providing the fundamental model of causality strikes many as rather naive. But it is not altogether implausible. It is in fact peculiarly congruent with the familiar admission that singular causal statements may be known to be true without knowledge of relevant covering laws. In what more convincing setting could the thesis be sustained than in that in which we ourselves deliberately and intentionally effect changes in our surroundings? So the advantages of the agency model deserve to be collected. The following claims are, then, favored by that model as opposed to the covering law model:

(a) Particular causal processes, notably those involving human actions, may be directly perceived or known to occur, without reference at all to covering laws;[16]

(b) Without denying that there are causal laws covering some particular causal processes, notably inanimate causal processes, the affirmation of a human action's causing an event does not as such entail that that process falls under any covering law.

It would appear that these claims may be greatly strengthened. Since we have already favored a heterodox view of causality, we may as well see how extreme an account may be plausibly advanced, without violating any known conceptual constraint. Certainly, we should want to see whether,

(c) For some subset of human actions, there is, in principle, no formulable causal law under which pertinent causal sequences can be explained.

Thus far, we have merely shown that (b) is a strongly reasonable claim; (c) if true, would require a fundamental shift in our theory of explanation – in fact, it would argue a distinction between the physical sciences and the "human studies" (*Naturwissenschaften* and *Geisteswissenschaften*) that would challenge the familiar assumptions of the unity of science. Also, *if* (c) obtained and if we conceded, along the lines sketched, that inanimate causal processes do entail covering laws, then it is reasonably clear that

(d) Actions satisfying (c) cannot be reduced to mere physical processes – for instance, along the lines of Herbert Feigl's familiar use of the notion "physical$_2$."[17]

Furthermore, the most extreme claim that we could hope to make, conceding (c) and (d) and affecting the issue of the unity of science, would be

(e) Actions satisfying (c) and (d) cannot be identified in the causal processes in which they enter, in a purely extensional way.

This goes entirely contrary to Davidson's neat distinction – shared by a great many – that, although contexts of causal explanation (being linguistic) behave intensionally, causal processes (obtaining in nature) behave extensionally. Arthur Danto speaks of the causal connection as "semi-intensional", but he means by that, precisely, that the causal process *is* open to explanation under a covering law.[18] But that is not the sense here required. The sense of (e) required is that human actions – some human actions at least – cannot be assigned a causal role except under some description, consistently with (b), that is, without reference to any causal law at all; and that, once assigned a causal role, they may be treated extensionally (shall we say, "semi-extensionally"?) in the sense that if an action causes some event or change of state, it does so under any description that picks out that very same action and event. Claims (a) through (e), therefore, constitute a very different picture of the causal nature of human agency from that which obtains in standard theories of physical causality. The striking thing is that, even at this stage of the argument, our set of claims enjoys a certain *prima facie* reasonableness.

Now, the key consideration regarding the causal explanation of human action is this: the paradigm of human action is free action, action performed in such a way that it is presupposed that the human agent is at the moment of acting capable of refraining from acting thus and capable of performing an alternative action.[19] It is precisely because a human being is, as already noted, a creature essentially capable of causal agency (1) and because human agency is such that agents can perceive or know that they and others produce effects in the causal way (2), that (3) the paradigms of human agency are free actions. How else should we understand human agency unless at least as including the deliberate or intentional production of certain effects? It is because human agents act *to* produce intended effects that we concede that, without reference to covering laws, they may perceive themselves and others succeeding in bringing such effects about and they may come to know that their intentional actions have such effects. This means, of course, that the admission of the causal efficacy of human actions is incompatible with the admission that all causal processes fall under covering laws, where covering laws are deterministic in the strictest sense. For the notion of free action supposes that, at any time *t* at which an agent acts of his own volition, by choice, deliberately, intentionally, for a purpose or the like, he is capable of refraining from that action or of performing another action of some sort: if determinism signifies that real possibility is exhausted by whatever is actual, then human freedom is incompatible with determinism.[20]

But the admission of free action has further consequences. For one thing, (4) human actions can be identified *initially* only in intentional terms, that is, in terms of the intentional states of the agent. (4) is misleading, however. It is *not* the case that every human action is or needs to be describable in terms of an agent's actual *intentions*; but human actions must be described *intentionally*. For example, a man may be doodling in a kind of trance or shifting from foot to foot without being aware of doing so: to characterize what he is doing as occurring "in a kind of trance", "without being aware of doing so" *is* to characterize (at least incipiently) what he is doing in intentional terms but not in terms of his actual intentions. This seems to be an empirical point. Yet Davidson maintains that "Action does require that what the agent does is intentional under some description, and this in turn requires, I think, that what the agent does is known to him under some description".[21] This either confuses the difference between intentional descriptions and descriptions in terms of an agent's intentions or else it begs the question about the individuation of actions – probably it does both.[22] But if an action must be initially identified in intentional terms, then it is impossible to identify a *free* action *initially* except under some cognitively favored description, that is, intensionally. For example, concede Davidson's case of one and the same action identified under four alternative descriptions: *flipping the light switch, turning on the light, illuminating the room, alerting a prowler.*[23] If the agent may be said to have acted freely *in* flipping the light switch or in turning on the light or in illuminating the room, he cannot be said – *initially at least* – to have acted freely *in* alerting a prowler who happened to be in the house. If what he did is one and the same action under the four descriptions proferred, then if he acted freely *in* what he did, that action, now described as alerting a prowler, was indeed a free action. He did *not*, by choice, act *to* alert a prowler; by choice, he acted *to* flip the light switch, *to* turn on the light, *to* illuminate the room. But *what* he did, *initially* identified under the description of choice or intended action, is (on pain of contradiction) a free action under any description that picks out the same action. So the admission of (4) entails that, (5) for free actions, particular actions can be identified initially only intensionally (partially but only partially the sense of [(e)] of a previous tally [p. 88]); hence that (6) free actions can be reidentified only on intensional grounds, that is, by reference to rules, institutions, conventional practices, traditions or the like. There are for example no compelling extensional grounds that would oblige us to regard what occurred under the four descriptions provided as one and the same *action*. For one thing, one might well hold, along the lines developed by Alvin Goldman, that there were four different actions, three "generated" in different ways from the original (probably "basic") action of flipping the light switch.[24] In that case, even if the agent could be said to have performed an action in alerting a prowler, even if he acted freely in flipping the light switch, he did not act freely in alerting the prowler. Alternatively, one may deny that the agent performed any action at all *in* alerting the prowler: alerting the prowler may (on a suitable theory) have been the mere *effect* of what he did in the way of performing a free action.

There is no settled way of individuating actions. Certainly, admitting

Goldman's strategy as a viable one, there is no nonquestionbegging way of individuating actions in terms of the spatiotemporal boundaries of purely physical events or their causal consequences. This of course goes against Davidson.[25] The important thing to bear in mind is simply that the admission of (5) and (6) does not entail that the causality of human agency cannot be treated extensionally; it entails only that (7) although causal contexts behave extensionally, the identification and reidentification of free actions *that* function causally cannot be managed in a purely extensional way. Doubtless, we may concede that constraints regarding the same spatiotemporal boundaries and the same effects serve as extensionally relevant limits *on* the reidentification of the same action; but it does not follow that human actions may be *initially* identified in terms of such constraints or *physical* constraints like them. Clearly, the expectation that it would follow is likely to be linked, as in Davidson's case,[26] to some form of reductionism.

Human actions, then, *qua* causes, may be said to behave, as all causal connections do, extensionally: whatever causes a given effect causes it. The trouble is that human actions, particularly free actions, cannot be identified or reidentified except in intensional terms. So the extensional feature of causality *is inoperative, where the explanation of free action is concerned, except under intensional identificatory constraints*. But if this is so and if causal agency does not entail covering laws (b), then we have very effectively begun to shape a defense for (c), that free actions cannot be explained under covering laws at all.

It is, in fact, the discovery of (7), that the identification and reidentification of free actions cannot be managed except intensionally, that must surely be the key to the often seriously marred insistence of so many post-Wittgensteinian theorists of action who deny that human action may be explained in causal terms.[27] They have, somehow, confused the issue of causal explanation with an essential condition on which human action may be causally explained. They were right to press the point because its admission enables us to grasp the very distinction of the causal explanation of human action; but they wrongly supposed that it actually precluded such explanation. The motivation is obvious: having assumed that causal explanation must involve covering laws and phenomena identifiable in a "purely extensional way",[28] perhaps even deterministic laws, the post-Wittgensteinians simply opted for construing the explanation of human action – particularly, free actions – as properly noncausal in nature. But if claims (a)–(e) were vindicated or some set of claims very much like them, then it would be possible to concede the conceptual insight of the post-Wittgensteinians – that the identification of human actions must be made in intentional terms, in particular, that certain actions, notably free actions, cannot be identified or reidentified except intensionally – at the same time we admit that human action is explainable in causal terms.[29]

On the other hand, emphasis on the problem of identification and reidentification shows both the implausibility of *identifying human actions with physical events* and the effect on the causal explanation of actions of rejecting such an identity. Consider that Davidson's favored specimen of a free action is a particularly simple one like flipping a light switch. Here, there *is* some temptation to identify human actions with physical events. In fact, Davidson asserts explicitly: "our primitive actions, the ones we do not by doing something else, mere movements of the body – these are all the actions there are. We never do more than move our bodies: the rest is up to nature".[30] The notion of a primitive action is closely related to that of a basic action as employed by Danto and Goldman: basic actions are taken to be identical with bodily movements – hence, to be identifiable in a purely extensional way (as physical events) and to permit the conventional covering-law explanation. Danto is particularly explicit about this.[31] But apart from the difficulty of specifying *which* actions are or must be basic actions,[32] there is no clear way of showing: either (i) that, for more complex actions – particularly those that have a large cultural import – actions can be traced to some putatively basic action that initiated the "generated" series; or (ii) that (in Davidson's sense) once we have accounted for how we move our bodies (as in primitive actions), the rest (presumably, whatever would causally explain such complex actions) is "up to [physical] nature"; or (iii) that the causal explanation of basic actions constitutes or entails a causal explanation of seemingly more complex, culturally qualified actions – which, on Davidson's view, would be identical with primitive actions and, on Goldman's view, would (in various ways) be "generated" from basic actions. For example, suppose it to be the case that one of President Nixon's speeches caused even his closest cohorts to press for his resignation. What *bodily movements* could serve as the basic actions with which his speaking could be identified? How could his action of uttering *significant words* be explained by an explanation of how he produced a certain sequence of *sounds*? Most important, how could the explanation of the effect, upon the culturally informed *decision* of his cohorts, of *his having spoken in the role of beleaguered President* possibly be reduced to the explanation of the bodily movements taken to be the basic actions underlying his speech?

There is no room here to attempt to refute the intended reductionism.[33]

But we must see that the viability of the explanatory claim about basic actions presupposes some form of mind/body reductionism. That thesis has not yet been satisfactorily formulated: in the case of actions of the complexity of Nixon's speaking, the weakness of the reductive program is peculiarly obvious.[34]

A very useful economy suggests itself, therefore, Let us leave to one side – as not proved – the identity thesis. The fact is that, even if it were valid, appeal to it would not clarify the way in which we explain human actions causally without some explicit strategy showing *how* to move from the explanation of intensionally identified actions (5) to the explanation of purely physical movements. Of course, *if* a relationship other than identity obtains between the mental or psychological and the physical, we may well explain the *bodily movements* that are, by that alternative relation, associated or linked with *human actions* without at all explaining the actions thus linked.[35] If we remain agnostic, therefore, about the relationship between the mental and the physical, we may at least regard it as questionbegging or unresponsive to claim that, in explaining the putative basic actions on which complex, culturally significant actions somehow depend, we are explaining the latter as well. There is no need to deny that the *bodily movements* (agnostically conceded to be) "linked" to human actions, particularly free actions, fall under covering causal laws in the manner already sketched; also, granting the linkage, there is no need to deny that, in some fair sense, the explanation of those bodily movements forms at least a *part* of the explanation of free actions. But the essential issue of explaining the causal role of intensionally qualified actions – and, correspondingly, of intensionally qualified effects – remains a mystery.

The point that is entirely missed by all those who anticipate reducing complex actions to basic or primitive actions (*à la* Davidson) or who anticipate linking them to the basic actions from which they are "generated", as far as providing an adequate causal explanation is concerned (*à la* Goldman), is simply that the explanation of free actions, cultural phenomena in general, tends not to focus at all or to emphasize the causation of the bodily movements with which particular actions and the like are linked. What causal account can be given of the production of Dante's *Commedia*? What explains the event of Hitler's rise to power? What explains the hysterical compulsion of Emma, in Freud's *Scientific Project*? Even the most fundamental actions that could be linked with these complex matters are never, in any obvious way, identified as basic bodily movements of any sort; the relevant actions (however "basic") are already richly informed by intentional considerations *that cannot be identified except in the context of an enveloping culture.* Broadly speaking, when we explain human actions, particularly free actions and phenomena identified as culturally important, we explain phenomena that are distinguished functionally. To explain a free action is to explain a functionally distinctive phenomenon, the function of which is specifiable in and only in a cultural context. So if (7) be admitted, that is, that free actions can be identified and reidentified only intensionally, it seems reasonable to insist as well that, (8) the causal explanation of free actions is essentially concerned with the explanation of functionally specified phenomena, where to specify an action functionally, that is, in culturally relevant terms, is to specify the intensional properties of that action. In short, to explain a free action is to explain a phenomenon that is not initially identifiable except in intensional terms *and* to explain the causal relations in which, *qua* possessing intensionally qualified intentional properties, it actually enters. So the cultural disciplines need never deny the relevance of causal accounts in purely physical terms . . . of, that is, bodily movements somehow "linked" to free actions or to related cultural phenomena. But *in* explaining such phenomena, those disciplines are primarily concerned to explain how it is, say, that a human being steeped in certain literary traditions, religious, philosophical, and political doctrines, produces a poem of such invention and significance as the *Commedia* exhibits; how it is that, in a certain historical setting in Germany, a moral monster was able to captivate an entire population and to move people to unheard-of acts of violence; how it is that, having suffered certain traumata at an early age, a young woman finds herself now unable to go into shops alone.

The reason the purely physical explanation of human actions is largely beside the point – though, of course, hardly irrelevant – is quite straightforward. In conceding that, (8) free actions and cultural phenomena are functionally distinguished, we implicitly concede as well that *there are indefinitely many alternative ways in which a culturally (or functionally) characterized action or phenomenon may be "linked" to some bodily event.*[36] Imagine for instance that Dante composed the *Commedia* by writing with his toes; or imagine that Hitler's recorded speeches were synchronized (undetected) with his (seemingly) live ranting. There is absolutely no reason to believe that, under such circumstances, *which* basic bodily movement happened to be the one by which the culturally significant actions of Dante and Hitler were manifested bears in any important explanatory way on the causal account of the latter. True enough, *some* account of physical causation must be relevant, that is, some account of how the action was actually "conveyed" or "manifested". But that tells us nothing of *what happened* in the pertinent sense (of what was thus conveyed); *and* substantial differences among possible physical movements suited to the phenomena in question are likely to have little or no bearing on the explanation of the particular phenomena themselves. Once conceded, the argument is a *reductio* of the basic-action model. For *if* free actions must be intensionally specified and if what is causally important in the explanation of free action bears directly on properties that can be specified only intensionally, then, without mounting a full-scale argument, we have good grounds for resisting any version of the identity thesis.

The point may be put more carefully. If free actions are functionally and intensionally specified, then there is no reason to suppose that whatever causal regularities they may be shown to exhibit (at that level at which they are first identified) could, in principle, be directly expressed in terms of coextensive lawlike regularities at the level of the physically specified phenomena with which (on the reductionist's view) they are identical. And if causal ascriptions at the level of human action presuppose reference to a holistic model of rationality – that is, that actions are themselves identified and treated intensionally – then no type identity thesis is tenable. Two possibilities have been pursued. One is that there *are* lawlike regularities at the level of functionally identified behavior and that these are empirically exhaustible by sets of disjunctive lawlike regularities confirmed at the physical level by means of entirely independent, extensionally manageable conditions of individuation.[37] The other is the view that, though they *do* play a causal role, there are *no* lawlike regularities at the level at which mental events and human actions are intensionally specified (and may be known to function causally).[38] Both have been offered as versions of *token* identity theories. But neither shows why identity should or must be supported at all. The second is explicitly incoherent because it does concede a causal role to mental events, a role empirically assignable at the level at which such events are identified as such, and does insist at the same time that causal contexts behave extensionally. The first is not explicitly incoherent because it does not address the question directly; but it offers no grounds for supposing that causal regularities at the "functional" level are lawlike, or that, if they are lawlike, there could be any disjunctive set of causal laws applied to extensionally sorted physical phenomena that would exhaust or tend to exhaust such regularities. There is, in short, every reason to believe that, at the complex cultural level at which, say, the various Cubists began to produce quantities of paintings both clearly convergent with respect to a developing tradition and also idiosyncratically distinctive, moving from artist to artist and from an earlier to a later phase in the work of a particular artist, the salient causal regularities in question would not describe lawlike or invariant uniformities but only uniformities *themselves contingently produced within this historical interval or that*; and that we could not expect to find explanatory principles or regularities of a comparable or reasonably suitable gauge that did not rely in a decisive way on such uniformities.

If this much be conceded, then, admitting (7) and (8), we have made a very strong gain on vindicating claim (c), that is, that for some subset of human actions, there is, in principle, no formulable causal law under which pertinent causal sequences can be explained. (c) does not follow, of course. But *if* we could strengthen the claim that the intensionally specified intentional properties of cultural life are precisely not the sort of thing that either could be convincingly treated as exhibiting lawlike regularities or must be so construed, we should have strengthened the reasonableness of (c) as well. The supporting grounds are, as it happens, hardly difficult to muster. First of all, authors like Herbert Feigl and Wilfrid Sellars regularly admit that intentional features (in Brentano's sense) are "irreducible to a physicalistic description".[39] Secondily, if we construe intentional features (now, including inten*s*ionally qualified intentional features) as the *actual*, distinguishing traits, properties, capacities of human persons, then the irreducibility conceded turns out (contrary to Feigl's and Sellars's purpose) to undermine the prospects of mind/body reductionism. Some treat inten*t*ionality entirely in terms of the inten*s*ional. But even if that claim were conceded (which is unconvincing[40] – at least for the reason that languageless animals appear to have mental states), the admission of linguistic *abilities* (as distinct from the abstraction, language) would restore our problem. If this much be conceded, then we may simply press the obvious: that there appear to be no laws of human history,[41] no laws of historical change as such. This is not to deny that there *may* be political, economic, sociological, psychological, psychopharmacological laws or laws of similar sorts. On the theory thus far advanced, such laws could not be more than statistical laws. But more than that, they would have to be laws that were intentionally informed in the peculiar way cultural phenomena are informed. What must be emphasized is that there need not be any causal laws of such sorts in order to provide an intelligible causal explanation of particular human actions (b); *and* that it may be quite impossible to formulate relevant covering laws for at least much that we are prepared to countenance as requiring causal explanation (c).

(c) would be strengthened if the *causal* explanation of human actions were, characteristically, or at least in a wide range of important cases, given in terms of historically relevant factors – factors that appear not to be of the sort for which covering laws may be discovered (for instance, where to hold that there were covering laws would entail that those laws were laws of historical change). Thus, to explain how it was that Hitler mobilized Germany in the way he did, one might have to resort to historically qualified sensibilities among the Germans – possibly along the lines by which Erich Fromm tries to show the sado-masochistic and dependency traits of just the sub-population that rallied to Hitler's call.[42] One could not hope to explain the transformation of capitalist Germany into the peculiar totalitarian form of the expanding Nazi empire by way of any laws of historical phases of development. And the explanation in terms of a combination of economic and psychological regularities, however reasonable, neither has the clear force of subsuming the phenomenon under a genuinely statistical law nor demonstrates that all cultural phenomena fall under covering laws.

The issue seems (but only seems) to be stalemated. The crucial feature affecting all such explanations is already in our hands. It is (7), that the identification and reidentification of free actions cannot be managed except intensionally. In short, *if* there are culturally pertinent (*geisteswissenschaftlichen*) laws, they would have to be formulated in terms of statistical regularities ranging over large sets of phenomena from different historical contexts. But that would mean: first, that a particular human action and the like would initially be able to be causally explained in terms of the intensionally qualified properties belonging to it and its peculiar cultural or historical milieu; and second, that the lawlike uniformities allegedly ranging over culturally or historically distinct sets of *such* phenomena would themselves require that functional similarities, *under some transculturally favored intensional qualification*, could be ascribed to *behavior thus intensionally qualified* (that is historically and culturally distinctive) within different societies. First, then (let us say), we explain how, in terms uniquely appropriate to the cultural life of Germany, it was that Hitler transformed the German people; *then*, assuming, say, that Hitler's ability and those of other historical figures constitute instances of a certain general charismatic power,[43] we attempt to formulate the *laws* of personal charisma.

What this shows is the reasonableness of (c), that is, that for some subset of human actions, there is, in principle, no formulable causal law under which pertinent causal sequences can be explained. The reason is a multiple one: first, because any arguably valid laws of the required sort both express and presuppose uniformities that cannot be collected in a "purely extensional way" (5, 6) – hence, that cannot be tested except *via* the intensional criteria by which they were first collected; second, because the phenomena in question are functionally specified – hence, that whatever is assigned a causal role in a "purely extensional way", bodily movements for instance, casts little or no light on the explanation of the culturally pertinent phenomena in question; and third, because the underlying phenomena on which the would-be culturally relevant laws depend are themselves first explained in terms of intensional distinctions prevailing in their own historical milieux. But, *if* culturally pertinent laws are logically weak in the manner marked, then there would appear to be no likelihood that the causal phenomena to be first explained *could*, as such, be subsumed under covering laws. But that, precisely, *is* claim (c).

In matters of this sort, we normally cannot hope to provide a knock-down argument. Still, the claim may be tightened further. What needs to be stressed is simply that the intensionally qualified uniformities that we appeal to in explaining human actions, particularly free actions and related cultural phenomena, are nothing more than the system of traditions, institutions, doctrines, practices, rules, culturally induced habits and the like that (i) psychologically inform in various ways the behavior of particular human agents belonging to a given culture; and (ii) *are themselves subject to causal change.* Traditions, institutions, and the rest, therefore, cannot be treated as or as directly yielding causal laws. Yet we do explain in the causal sense particular actions and the particular consequences of human action by reference to such general but severely circumscribed regularities. In fact, the very idea that a human agent, performing a free act, acts intentionally – in accord with a purpose or plan of action, by choice, on the basis of appraising available alternatives – makes no sense unless the intentional properties that inform his behavior are suitably groomed by the organized and organizing cultural milieu in which he develops. Hence: to be a human agent is to behave in ways explainable in the causal sense by reference to the intensionally qualified institutions of one's own culture. Since, as has been said, such institutions are themselves subject to causal change (through the inventive agency of human beings), the relevant form of explanation – what we may call explanation under *covering institutions* – cannot be subsumed under the model of explanation by covering law. (c) therefore, seems to be true, without endangering the prospect or intelligibility of causal explanation itself.

NOTES

[1] Ernest Nagel, *The Structure of Science* (New York: Harcourt, Brace and World, 1961), Chapter IV.
[2] *Ibid.*, p. 74.
[3] Carl G. Hempel, 'The Logic of Functional Analysis', in *Aspects of Scientific Explanation* (New York: Free Press, 1965), pp. 300–301.
[4] *Ibid.*, p. 299.
[5] Wesley C. Salmon, 'Statistical Explanation', in Wesley C. Salmon *et al.*, *Statistical Explanation & Statistical Relevance* (Pittsburgh: University of Pittsburgh Press, 1971), pp. 71–72.
[6] *Ibid.*, pp. 64–65.
[7] *Ibid.*, p. 43.
[8] Hans Reichenbach, *The Direction of Time* (Berkeley and Los Angeles: University of California Press, 1956).
[9] Salmon, *op. cit.*, p. 68.
[10] Cf. R. S. Peters, *The Concept of Motivation* (London: Routledge and Kegan Paul, 1958).
[11] Donald Davidson, 'Causal Relations', *Journal of Philosophy*, LXIV (1967), 700–701. For a scrupulous account of Hume and Mill on causality, see Tom L. Beauchamp and Alexander Rosenberg, *Hume and the Problem of Causation* (New York: Oxford University Press, 1981), especially Chapter 8. Cf. also, Barry Stroud, *Hume* (London: Routledge and Kegan Paul, 1977), Chapters 3–4.
[12] Cf. Donald Davidson, 'Mental Events', in Lawrence Foster and J. W. Swanson (eds.), *Experience & Theory* (Amherst: University of Massachusetts Press, 1970).
[13] J. J. C. Smart, *Philosophy and Scientific Realism* (London: Routledge and Kegan Paul, 1963), Chapter 3.
[14] Karl R. Popper, 'The Aim of Science', in *Objective Knowledge* (Oxford: Clarendon, 1972), pp. 196–197.
[15] *Ibid.*, p. 197.
[16] Cf. Curt J. Ducasse, 'Critique of Hume's Conception of Causality', *Journal of Phiilosophy*, LXIII (1966).
[17] Cf. Herbert Feigl, *The 'Mental' and the 'Physical'. The Essay and a Postscript* (Minneapolis: University of Minnesota Press, 1958, 1967); also, Joseph Margolis, *Persons and Minds* (Dordrecht: D. Reidel, 1978).
[18] Arthur C. Danto, *An Analytical Philosophy of Action* (Cambridge: Cambridge University Press, 1973), p. 98.
[19] One of the most perceptive accounts of this aspect of human action appears in J. L. Austin, 'Ifs and Cans', *Philosophical Papers* (Oxford: Clarendon, 1961).
[20] Cf. Roderick M. Chisholm, 'Human Freedom and the Self', *Freedom & Morality*, John Bricke (ed.) (Lawrence: University of Kansas Press, 1976).
[21] Donald Davidson, 'Agency', in Robert Binkley, Richard Bronaugh, and Ansonio Marras (eds.), *Agent, Action, and Reason* (Toronto: University of Toronto Press, 1971), p. 12.
[22] See 4, above.
[23] Donald Davidson, 'Actions, Reasons and Causes', *Journal of Philosophy*, LX (1963). On the issue of explanation and identity of actions, see Joseph Margolis, 'Puzzles about

Explanations by Reasons and Explanation by Causes', *Journal of Philosophy*, LXVII (1970); review of Alvin Goldman, *A Theory of Action, Metaphilosophy*, V (1974).
[24] Alvin Goldman, *A Theory of Action* (Englewood Cliffs, N.J.: Prentice-Hall, 1970).
[25] Cf. Donald Davidson, 'The Individuation of Events', in Nicholas Rescher *et al*. (eds.), *Essays in Honor of Carl G. Hempel* (Dordrecht: D. Reidel, 1969); Margolis, *Persons and Minds*, Chapter 13.
[26] Cf. 'Mental Events.'
[27] Cf. for instance A. I. Melden, *Free Action* (London: Routledge and Kegan Paul, 1961).
[28] Davidson, Agency', p. 8.
[29] Cf. 4, above.
[30] 'Agency', p. 23.
[31] Cf. Danto, *op. cit*., pp. 97–115; also, Goldman, *loc. cit.*
[32] Cf. Myles Brand, 'Danto on Basic Actions', *Nous*, II (1968); and Frederick Stoutland, 'Basic Actions and Causality', *Journal of Philosophy*, LXV (1968).
[33] Cf. *Persons and Minds.*
[34] Davidson does of course subscribe to a version of the identity theory; but it is fair to say that the thesis is merely announced, not at all defended. Cf. 'Mental Events'; for a sustained criticism of Davidson's view, see Joseph Margolis, 'Prospects for an Extensionalist Psychology of Action', *Journal for the Theory of Social Behavior*, XI (1981).
[35] See *Persons and Minds.*
[36] Cf. Hilary Putnam, 'Minds and Machines', in Sidney Hook (ed.), *Dimensions of Mind* (New York: New York University Press, 1960); and Jerry A. Fodor, *Psychological Explanation* (New York: Random House, 1968).
[37] This view is favored by Jerry Fodor, *op. cit*., pp. 9–26.
[38] This view is favored by Donald Davidson (in 'Mental Events') – the position he dubs "anomalous monism".
[39] Feigl, *op. cit*., pp. 50–51. Cf. Wilfrid Sellars, 'Philosophy and the Scientific Image of Man', in *Science, Perception, and Reality* (London: Routledge and Kegan Paul, 1963); and 'A Semantical Solution of the Mind-Body Problem', *Methodos*, V (1953).
[40] Cf. James Cornman, 'Intentionality and Intensionality', *Philosophical Quarterly*, XII (1962).
[41] Cf. Karl R. Popper, *The Poverty of Historicism* (London: Routledge and Kegan Paul, 1957).
[42] Erich Fromm, *Escape from Freedom* (New York: Rinehart, 1947).
[43] Cf. H. H. Gerth and C. Wright Mills (trans. and eds.), *From Max Weber: Essays in Sociology* (New York: Oxford University Press), Chapters 9–10.

6

COGNITIVISM AND THE PROBLEM OF EXPLAINING HUMAN INTELLIGENCE

The explanation of intelligent behavior is an issue of the greatest strategic importance in any attempt to understand the conceptual features of the psychological, social, and cultural sciences. It is, however, too global an issue to be usefully confronted without substantial constraints. For example, a sensible beginning suggests that analysis should be at least initially restricted to the linguistically informed behavior of human beings, avoiding generalizations ranging over nonlanguage-using animals. In any case, it may be argued that the study of animal intelligence is conceptually dependent on the use of categories paradigmatically provided for the study of human intelligence. This, of course, is not to say that only humans *describe* animal intelligence. It is to say (rather) that animal intelligence is *modelled* on the human, in the sense that intelligence entails the ascription of propositional attitudes and that the structure of the propositional content of such attitudes is, and must be, modelled on sentences. Similarly, the issue may be fairly freed from the question of physicalistic reduction, in the plain sense that, should reductionism obtain, the phenomena of human intelligence would remain psychologically real, and the very success of any reductive effort would initially concede the apparent distinction of the phenomena thus reduced. The prospect of similar economies invites us to attempt to isolate what, regarding the explanation of human intelligence, may most decisively affect our assessment of the methodological import of the question.

Certainly, one particularly strategic aspect of the issue concerns whether, in explaining intelligent behavior, it is either necessarily true, or at least tenable to hold, that relevant and adequate explanations may be formulated entirely in terms of internal, sub-molar processes that are themselves at least provisionally characterized as cognitive or rational or intelligent – that is, as involving homuncular processes of some sort – or as molar processes not accessible or not fully accessible at the conscious level. Such a program of explanation is what is generally termed cognitivism.[1] Homuncular cognitivism is perhaps most explicitly championed by Daniel Dennett;[2] and molar cognitivism – in at least one extreme and memorable form – by Jerry Fodor.[3] But Polanyi probably offers an informal version of molar cognitivism, at least with respect to his doctrine of "tacit knowledge",[4] with which Merleau-Ponty's views noticeably converge.[5] And both Chomsky and Piaget, who rather strenuously oppose one another,[6] are at least partially attracted to forms of molar cognitivism; they demur (in different ways) because they are also attracted to doctrines regarding the biological development of cognitive abilities that preclude explanation of such development in terms of exercising cognitive powers already in place.[7] In general, cognitivism postulates infra-psychological computational powers that explain intelligent behavior (or "output") as the result of the internal processing of contingent data (or "in-put").[8] In principle, cognitivism need not be restricted to linguistic behavior – need not even be restricted to behavior that is linguistically informed. It may be applied to sensory perception among animals, for instance.

Now, homuncular explanations can be shown to be dependent, in principle, on molar explanations of intelligence, simply because intentionally specified homunculi are themselves merely the *relationally* specified sub-functions of the molar functions of an intelligent system. This is not to deny a role to homuncular explanation, only to confirm that it can never replace, without remainder at the molar level, molar explanations themselves. Consequently, the defeat of homuncular cognitivism hardly touches the prospects of molar cognitivism. But it does show that homuncular cognitivism cannot provide an intermediary step for any purely physicalistic or extensionalist reduction of intelligence that could not be independently supported.[9]

We are not, however, exclusively concerned with the adequacy of cognitivism. The larger biological models that Chomsky and Piaget favor, for instance, suggest the advantage of casting a wider net. We are looking for certain common features of a range of theories regarding the explanation of intelligent behavior that share with cognitivism a certain orientation which – once shown to be justified or unjustified or doubtful – would decisively affect the development of all those disciplines focussed on the details and causal processes of human existence. The issue has not been very thoroughly canvassed. In fact, it may well be instructive to bring our question to bear on a variety of theories not often linked to one another. Our general strategy will be to explore in a dialectical way three fundamentally distinct approaches to human intelligence – those that favor internal computational processes or full cognitive competence (I), those that favor ecological niches and (at least chiefly) some sort of noncognitive symbiosis between organism and environment (II), and those that favor initiation into a contingent culture resulting in at least a moderate measure of cognitive control (III). The argument here advanced inclines (nonexclusively) toward (III); that is, there can, on (III), only be a limited role assigned, in explaining human intelligence, to theories of types (I) and (II), but such theories can be accommodated.

I

Consider, first, H. P. Grice's theory of (nonnatural – essentially, linguistic) meaning:

> "U meant something by uttering x" is true if, for some audience A, U uttered x intending
> (1) A to produce a particular response *r*.
> (2) A to think (recognize) that U intends (1).
> (3) A to fulfill (1) on the basis of his fulfillment of (2).[10]

The adequacy of Grice's formulation need not concern us; whatever adjustments Grice is willing to consider, the central thesis – that of the dependence of the meaning of a linguistic utterance, and of the conditions of its being understood, on a speaker's intentions – has remained basically unaltered. Furthermore, the very formula cited makes it reasonably clear that, on Grice's view, U may be said to have "meant something" only if U consciously or deliberately or intentionally (or the like) uttered what he did with the indicated (rather complex) objective in mind – or at least with something very much like it in mind. In effect, then, Grice is attracted to something very much like a molar analogue of the cognitivist theory of meaning, though he does not explore the issue in this respect.[11]

Nevertheless, Grice does appear to hold that, in the normal context in which speakers mean what they say, the utterer has in mind, in a psychologically real sense, a certain complex intention, grasping which the hearer may reasonably be taken to understand what is said. Meaning, therefore, is primarily a matter of real psychological states. If, for instance, one thought that what a speaker said could be assigned its normal meaning in a way that did not require anything like Grice's complex intention (as in a naive speaker's hitting on a locution that he doesn't really understand but that is meaningful enough), or in a way that could not be restricted to his intention or did not depend essentially or primarily on his intention, even if that were somewhat congruent with what was said (as in supposing that the regularities of a language may obtain without being in any sense completely internalized psychologically in any speaker), then Grice's approach to the analysis of meaning would be fundamentally mistaken. In short, the intentional regularities of language might be socially real – actually shared by a society in some sense – without being distributively internalized in the mental states of all participating speakers. It would then still be possible to agree with Grice that, at least initially, the analysis of language should concede (without prejudice to extensional claims) that "intensionality seems to be embedded in the very foundations of the theory of language"[12] – without also holding (with Grice) that, again initially, the analysis of meaning should be formulated ***entirely*** (or even primarily) in terms of the speaker's psychologically real *intentions.*[13] In fact, this way of putting matters shows the ease with which the distinction (introduced earlier) between ontic and attribute dualism can be used to accommodate the adoption of methodological individualism – in the social sciences – without denying the reality of social relations and attributes of social aggregates not reducible to sets of attributes ascribable to individual human agents. Only human individuals actually speak, but languages are analyzable only in terms of attributes ascribed to societies or social aggregates.

The plausibility of resisting Grice's view becomes quite clear once we concede that, in a complex linguistic society, it is by no means obvious that the intelligibility of written language is merely or entirely parasitic on direct speaker-hearer encounters and once we concede that the very intelligibility of what is said in oral exchange is bound to be dependent on the complex practices of a society that no competent speaker could be entirely conscious of, or could somehow have successfully internalized psychologically. Certainly, this will be plain enough regarding young children moving within the patterns of adult discourse; and if conceded there, the point will hardly be resisted for adult discourse itself. Concede, for example, that the *meaning* of what some speaker or writer says is, at least in part, dependent on the nonlinguistic experience of an historically changing society, or conveyed, as if by a kind of division of (habituated) labor, by assuming an accord with the variably specialized experience and practice of quite different (and somewhat segregated) sub-populations within one's own society:[14] Grice's over-simplification would then begin to show itself.

There are, in fact, at least three rather prominent doctrines that have been historically linked with the development of current theories of language that decisively bear on the resolution of our question regarding the explanation of intelligent behavior – that Grice's theory rather weakly touches on. These may be suggested by the following statements. The first is by Louis Hjelmslev:

> *A priori* it would seem to be a generally valid thesis that for every *process* there is a corresponding *system*, by which the process can be analyzed and described by means of a limited number of premisses. It must be assumed that any process can be analyzed

into a limited number of elements recurring in various combinations. Then, on the basis of this analysis, it should be possible to order these elements into classes according to their possibilities of combination. And it should be further possible to set up a general and exhaustive calculus of the possible combinations. A history so established should rise above the level of mere primitive description to that of a systematic, exact, and generalizing science, in the theory of which all events (possible combinations of elements) are foreseen and the conditions for their realization established.[15]

The second, by Ferdinand de Saussure:

If we could embrace the sum of word-images stored in the minds of all individuals, we could identify the social bond that constitutes language [*la langue*]. It is a storehouse filled by the members of a given community through their active use of speaking [*la parole*], a grammatical system that has a potential existence in each brain, or more specifically, in the brains of a group of individuals. For language is not complete in any speaker; it exists perfectly only within a collectivity .... It is the social side of speech, outside the individual who can never create nor modify it by himself; it exists only by virtue of a sort of contract signed by the members of a community. Moreover, the individual must always serve an apprenticeship in order to learn the functioning of language; a child assimiliates it only gradually.[16]

The third, by Noam Chomsky:

... linguistic theory (or "universal grammar") is what we may suppose to be biologically given, a genetically determined property of the species: the child does not learn this theory, but rather applies it in developing knowledge of language .... [But if] non-linguistic factors [must be included in] grammar [in the manner of the generative semanticists]: beliefs, attitudes, etc., [this would] amount to a rejection of the initial idealization to language, as an object of study. A priori, such a move cannot be ruled out, but it must be empirically motivated. If it proves to be correct, I would conclude that language is a chaos that is not worth studying .... Note that the question is not whether belief or attitudes, and so on, play a role in linguistic behavior or linguistic judgments .... The question is whether distinct cognitive structures can be identified, which interact in the real use of language and linguistic judgments, the grammatical system being one of these.[17]

Hjelmslev's concept of a theory is notably reminiscent of Wittgenstein's project in the *Tractatus*. On Hjelmslev's view, "a theory ... is in itself independent of any experience. In itself, it says nothing at all about the possibility of its application and relation to empirical data." It is, in this sense (he says), "arbitrary"; and, although it is designed to assign an "underlying" system to (linguistic) processes (in the case in question), it functions primarily to provide a deductive system of sufficient power to represent all possible states of affairs.[18] In this sense, of course, it serves as a remarkably clear specimen of so-called structuralist notions of explanation. On the other hand, Hjelmslev wishes to constrain his theories so that they function in an empirically pertinent and testable way; they must, in other words, be "appropriate". And, in being confirmed as appropriate, in identifying the underlying system of linguistic processes, they may be said to be psychologically "realized".[19] Since the elements of Hjelmslev's system are defined relationally, and since the governing theory invoked is itself "arbitrary" and only one among alternative such systems, Hjelmslev's account provides somewhat indifferently for what are often called "heuristic" as opposed to "realistic" explanations; explanatory "adequacy" is to be preferred to any form of "naive realism".[20] Not unreasonably, one could then construe Hjelmslev's dictum – that to every process there corresponds an underlying system – as affirming no more than the truism that every finite natural process may be finitely analyzed into elements and its occurrence explained as the ordered combination of those elements, without regard to realistic considerations. This would correspond rather closely to the nonstructuralist-motivated observation of Hilary Putnam's, that "everything is a Probabilistic Automaton under some Description".[21] But the nature of Hjelmslev's venture makes it clear as well that he himself supposed that a comprehensive system could (nonexclusively) be postulated for the entire range of real linguistic processes, that it would include a finite set of elements and (in effect) a finite set of rules for recursive combination, and that it could be reasonably confirmed as such in an empirical manner. In fact, Hjelmslev's proposal suggests how one may invoke a cognitivist or computational model ***heuristically, consistent with a realism regarding actual speech***, thereby avoiding the typical platonism of realistically construed computational models (notably, Jerry Fodor's).[22] For, *if* elements of a particular computational model of learning and cognitive competence are themselves construed realistically – inevitably, innately, in order to initiate any pertinent process – then it will become impossible to account for the "learning" of "new" concepts, say, except in terms of recursive and other combinatorial possibilities regarding some original, finite, and completely adequate set of such elements. No remotely promising set of such elements has ever been actually specified.

So seen, Hjelmslev's account is noticeably restricted or deficient in a number of ways that bear on our issue. For one thing, Hjelmslev is not particularly concerned, in the manner of the cognitivists, to establish his account of linguistic production *as* psychologically real rather than as explanatorily adequate to real natural languages; in fact, he treats language as a *sui generis* system to be analyzed independently of psychological and neurophysiological processes. Secondly, even if a system (in Hjelmslev's sense) could be constructed along structuralist lines, there would be no antecedent reason for supposing that if it could be used heuristically, it might also prove to be psychologically real; but if that issue were suitably pressed, then the very meaning of such a system's "underlying" linguistic processes would be open to an important dispute – trivially, it could be said to underlie such processes in a "structuralist" sense but not, perhaps, in any internal psychological or social or otherwise realist sense.[23] In any event, Hjelmslev makes no provision for resolving the issue. Thirdly, it is not at all clear that Hjelmslev *is* correct in supposing (in his own sense) that there must be a system underlying every process, if every such system must be of a kind that *could* actually generate such processes in a cognitively real sense. The matter has been disputed with respect to natural languages. For instance, in developing his well-known semantic conception of truth. Tarski provides in effect for a *sui generis* "generative" account that could be said to underlie a language, although he is quite explicit that his program applies only to "those languages whose structure has been exactly specified" in the relevant formal respects. Tarski holds quite straightforwardly that its application to natural languages could be only "approximate" at best, that it would depend on "replacing a natural language (or a portion of it in which we are interested) by one whose structure is exactly specified, [by one] which diverges from the given language 'as little as possible'". Tarski was not in the least sanguine about such application, contrary to the convictions of certain of his followers.[24] In any case, Tarski was not in the least concerned to explore the prospects of construing his theory as computationally real in the psychological sense. He certainly does not preclude such a possibility, but neither does he insist that such a system *must* obtain in the processes of natural language. Michael Arbib and David Caplan have recently insisted that "neurolinguistics must be computational,"[25] but they offer no argument to this effect, concede that it is an empirical matter, and even admit that it may well be false. Possibly the most extreme attempt to defend the general cognitivist thesis has been made by Jerry Fodor.[26] But Fodor's argument is so radically platonistic with respect to concepts (that is, that all concepts must be combinations of some real, innate set of elemental concepts) that it is empirically both incapable of actual specification (in terms of an adequate set of original concepts) and utterly implausible (in precluding any form of concept acquisition – possibly biological, possibly cultural – not reducible to the combinatorial powers of the original set).[27] Furthermore, the provision of a *system* (in Hjelmslev's sense, or in Putnam's, with respect to construing behavior and the processing of information on the model of automata) empirically applies, first, to finite *segments* of a given domain of intelligent life;

nothing regarding this entails any assurance that the system in question could accommodate the total *capacity* of the domain (that is, a capacity not antecedently known, as in the case of invented machines, and a capacity normally not completely exhibited in the range of behavior and processing within any manifest segment). The thesis that, under conditions of real-time constraint and historical contingency, an inference from the one to the other *is* empirically available is both utterly implausible and what, in the Continental idiom, is contemptuously termed "totalizing".

Saussure's thesis (to return to our specimen views) construes the underlying system of language (*la langue*) as (in some respect) real, but certainly not psychologically real – in the plain sense that no particular person *could* internalize the entire system in his speech (*la parole*); that, in speaking, no person need have internalized the entire system relevant to the production of admissible particular utterances; and that the system is real only in the sense that it ("structurally") underlies the aggregated speech of the members of a society, without its total structure ever being actually instantiated by any particular aggregate of speech acts. But even these *caveats* are insufficient to fix the principal sense in which Saussure is willing to postulate an underlying system, for they fail to capture his notion of the distinction between synchronic and diachronic linguistics. From an actual speaker's point of view, Saussure holds, diachrony is irrelevant; the system he realizes in his speech is "in itself ... unchangeable".[28] Nevertheless, Saussure admits that the elements of the system *are*, diachronically, "altered without regard for the solidarity that binds them to the whole [system]".[29] Hence, he remarks, we must concede "the ever *fortuitous* nature of a state [that is, roughly, a part of actual *parole* synchronically construed] .... [L]anguage is not a mechanism created and arranged with a view to the concepts to be expressed .... The diachronic perspective deals with phenomena that are unrelated to systems although they do condition them."[30]

On Saussure's view, "Neither [is] the whole [system] replaced nor [does] one system engender another; one element in the first system [is] changed [in the sense that *parole* is actually diachronic], and this change [is] enough to give rise to another system."[31] His point appears to be: that speech is fundamentally socialized though never completely internalized psychologically; that it works efficiently in spite of this; that it is open to diachronic changes and influences without failing to be communicatively effective; and that it (*parole*) is unformalizable in the sense in which language (*langue*) can be (serially) formalized.[32] Furthermore, insofar as *parole* is what it is, in virtue of instantiating some combination of the elements of *langue*,[33] and insofar as *langue* is some synchronic idealization fitted to an aggregate of actual linguistic events (*parole*), it is extremely difficult to say what the sense is in which *langue* is real. Even the elements of spoken or uttered language are defined relationally with respect to some synchronic system; they are not in any sense discrete elements (the actual events of *parole*) identified independently of such a system. Predictably, the sense in which *langue* is real is not very distant from Hjelmslev's (who, of course, was strongly influenced by Saussure). Accordingly, Saussure explicitly acknowledges that "the division of words into substantives, verbs, adjectives, etc. is not an undeniable linguistic reality": "Linguistics . . . works continuously with concepts forged by grammarians without knowing whether or not the concepts actually correspond to the constituents of the system of language."[34]

For our present purpose, it is perhaps sufficient to note that, on Saussure's view, the processes of language *are not, need not be, and even cannot be* computationally real, in the psychological sense the cognitivists have insisted on. And yet, in spite of that, Saussure is prepared to pursue the enterprise of articulating a set of systems in terms of which alone speech is rendered intelligible, any one of which, on the cognitivist view, might well have been computationally real. It is not that Saussure does not commit himself on the question. On the contrary, he does; and on his view, the dialectical nature of the relationship between *langue* and *parole*, the inevitable idealization of synchronic systems, the relational nature of the elements of language relative to some synchronic grammar, and the intrusion of diachronic forces affecting the speech of actual human agents, all bear decisively on the impossibility of there being a real, fixed (synchronic) grammar capable of generating the diachronic *passage* of all actual speech events.

There are palpable affinities, here, to the various – admittedly diverse and by no means equally defensible – theories advanced by such authors as the Wittgenstein of the *Philosophical Investigations*, Merleau-Ponty,[35] Pierre Bourdieu,[36] Polanyi,[37] and Thomas Kuhn.[38] The essential difference is that Saussure is a classic structuralist and the others are not. But that makes the comparison all the more interesting, since Saussure's synchronic systems are (in the cognitivist sense) incapable of generating the actual and psychologically possible utterances of a real language – though they purport to do so "structurally". The others also focus on the cognitive abilities of humans – in perception, intentional behavior, social practices, skills, interpretation, scientific inquiry, as well as in natural languages in general – but (in various ways) they deny the possibility of a computational algorithm or of any heuristic approximation to one that could be realistically construed as having actually generated intelligent human behavior.[39]

Noam Chomsky's thesis (as cited) is both curious and bold, viewed against the backdrop of the structuralist conception of language. For, in treating his linguistic universals as "species-specific", Chomsky clearly construes them as governing all possible natural languages; and yet, specifically in opposition to the structuralist attitude, he regards his universal grammar as biologically and psychologically real – "innate interpretive principles, . . . concepts that proceed from 'the power of understanding' itself, from the faculty of thinking rather than from external objects directly".[40] In effect, in revising his account of universal grammar under the pressure of continuing empirical studies of natural language, Chomsky appears to conform to the Saussurian effort to accommodate both the synchronic and diachronic aspects of linguistics; but his motivation remains more like that of Hjelmslev (methodologically though not metaphysically), in anticipating a comprehensive theory that need not attend at all to the historical dimension of natural languages – even if the discovery of that theory is a function of historical studies. It is, of course, precisely this insistence that there must be one fixed and omnicompetent set of rules adequate to determine (and, therefore, to understand) the potentially infinite supply of sentences of any and every natural language that was opposed by the post-structuralist literary and linguistic theorists (working chiefly in France) – who, in effect, thereby assimilated Chomsky to the structuralism they opposed.[41]

But the assimilation was a mistake. In resembling the structuralists in the respect sketched, Chomsky was actually diametrically opposed to their undertaking – in the precise sense that they characteristically introduced *their* ("arbitrary") abstract systems in order to facilitate our understanding the *historical* dimension of human language and society, whereas Chomsky introduces *his* in order to provide an *ahistorical* account of how real universals generate the sentences of actual languages (and of other structured forms of knowledge and intelligence as well).[42] Chomsky is quite explicit about the Humboldtian nature of his undertaking:

> It seems clear that we must regard linguistic competence – knowledge of a language – as an abstract system underlying behavior, a system constituted by rules that interact to determine the form and intrinsic meaning of a potentially infinite number of sentences. Such a system – a generative grammar – . . . defines a language in the Humboldtian sense, namely as 'a recursively generated system, where the laws of generation are fixed and invariant, but the scope and the specific manner in which they are applied remain entirely unspecified'.[43]

Hjelmslev's view, as we have seen, traded the universality of the theories he favored for a decisive blurring of the distinction between a heuristic and a realistic reading of grammatical structures. Both Hjelmslev and Saussure construed their systems as inherently abstract – even "arbitrary". But Chomsky rejects (Continental) structuralism in precisely the same sense in which he rejects the machine simulation of human intelligence: it is purely functionalist in nature, *too* abstract; in fact, he judged structuralism to have compounded the behaviorist's error, in attempting to imitate the linguistic behavior of humans.[44] Chomsky rejects the search for a discovery procedure accessible at the surface of actual speech, and *his* universals preserve the range of the Continental systems *at the same time they are counted as biologically and psychologically real* – incarnate, so to say, in some genetically prepared substrate of human existence. In this sense, Chomsky favors a particularly bold version of the cognitivist thesis, while he remains profoundly skeptical of machine simulations of the operations of transformational grammar.[45]

We are now in a position to appreciate the grave difficulties confronting all forms of the cognitivist conception of human intelligence. For one thing, *if* (as Saussure affirms and Chomsky denies) the generative structures of a language must be empirically learned from infancy, then it is impossible to defend a thoroughly cognitivist account of linguistic intelligence: no agent can have internalized psychologically all the rules requisite to the computation of admissible utterances. Also, the synchronic system of rules assigned to any range of actual *parole* cannot claim the universal scope Chomsky claims. Furthermore, the acquisition and possession of the deep rules of any language cannot be meaningfully imputed to any aggregate of human individuals. The rules belong to a social "collectivity" only, and must be constantly formulated under the pressure of diachronic changes in *parole*. Chomsky was quite right to see that *if* a universal grammar was to be defended *realistically*, it had to be defended on innatist grounds. This also explains the motivation for Fodor's otherwise extraordinary platonism with respect to semantic concepts.[46] But Chomsky does not directly discuss the *psychological* relationship between the use of innate rules and the use of the regularities of contingently acquired natural languages.

Furthermore, *if* (as Chomsky affirms) universals must be defended empirically – *not* in the structuralist manner – and *if* grammar cannot be specified in principle independently of semantic, referential, and nonlinguistic factors (independently of beliefs, attitudes, accumulating experience), then the innatist hypothesis *cannot* be defended (unless, *per impossibile*, all such factors are themselves internalized in the same way). This consideration confirms the initial plausibility of Saussure's emphasis on the dialectical relationship between *langue* and *parole*: the idealized structure of language must be made to fit the aggregated patterns of the speech of diachronically varying speakers. Hence, any synchronic system is conceptually linked to the linguistic behavior of an actual community; and the linguistic behavior of particular speakers is characterizable as such only in terms of an imputed system shared by some synchronically specified community. Chomsky not only treats his system of universals in a radically ahistorical manner but, by construing a universal grammar as fixed in a real psychological way independently of semantic, referential, and nonlinguistic factors, he somehow manages to separate the deep grammar of all historically formed, natural languages from those very factors with which they appear to be dialectically so intimately linked. How, we may ask, is it possible that there should be a grammar underlying all culturally formed languages that, apart from their very formation (really independent of, and prior to, their formation), is already biologically fixed? There is a fair sense in which merely to postulate such detached universals as (somehow) genetically fixed (and innately "known") is to advance a thesis that cannot be coherently reconciled with the social nature of human cognition – in particular, of linguistic acquisition – as well as with the empirically reasonable thesis that, in acquiring a (contingent) natural language, one acquires the grammar that that language entails. The conceptual dependence of a natural language on a deep grammar cannot, by itself, rightly justify the temporal priority of the latter's possession.[47]

This is not to say that Chomsky's claim cannot be redeemed. But there are further constraints that such a system would have to satisfy that are not clearly satisfied. In particular, the putative universals would have to be of a nature that *could* be genetically fixed; the likelihood that there were any such rules would diminish proportionately to the extent to which rules could be shown to be inextricably linked with the very factors Chomsky insists they are independent of. Initially, Chomsky held that the proposed linguistic universals could be fitted to – and would be empirically confirmed in virtue of fitting – the actual linguistic behavior of aggregates of human speakers; but gradually, the alleged universals became increasingly inexplicit, so that the conceptual need to admit some such innate system (to which particular schemata might approximate) took precedence over the empirical standing of any particular claim.[48] The innatist theory, therefore, is meant as an inference to the best explanation of the global phenomenon of language construed as an ahistorical system more than as an inference to the best explanation of the detailed features of actual human speech within the contingencies of particular cultures. Hence, in spite of his biological orientation, Chomsky's views have in a curious way a conceptual motivation rather closer to Saussure's than he would care to acknowledge.

Chomsky, of course, was actually swayed for a time by the development of the so-called generative semantics movement.[49] In any case, that movement poses a conditional dilemma for Chomsky's account. Either the surface structure of natural languages decomposes into a deep structure that is jointly semantic and syntactic (that requires, therefore, an ultimate set of semantic primitives), or else the actual form of any putatively deep structure is itself causally determined by the surface structure of any given – contingently acquired – language. The first threatens an implausibly rich platonism, deficient to the extent that a viable set of semantic primitives adequate to account for the complex system of concepts of any natural language seems to be beyond our reach. The second increases the vulnerability of any universalist claims *and* requires supplemental linguistic rules that must somehow be learned or acquired. Hence, a cognitivism based on Chomsky's theory is bound to run into serious difficulties of just the sort that Saussure's alternative approach attempted to accommodate – in particular, regarding the social nature of language and language acquisition and the intrusion of diachrony in the linguistic patterns of particular speakers. Chomsky himself quite frankly admits the uncertainty of invariance constraints on word order[50] and the like, on which, presumably, most universals depend. It may even be that universal regularities are not only idealizations that mask exceptions but (following Saussure) distorting abstractions – artifacts of the explanatory theory – that impute uniform processes to complex systems that bear to one another only convergent resemblances (for example, regarding the occurrence of sentences, noun phrase/verb phrase elements, possibly even subject/predicate structures).

II

We need, now, a brief review of some salient possibilities regarding the explanation of intelligent behavior, that would obviate cognitivism. Very probably, the most extreme model is the one offered by J. J. Gibson in his "ecological" approach to sensory perception.[51] Gibson rejects all forms of "the stimulus-response formula" and of empirical learning modelled on versions of that formula;[52] he also rejects all computational models focused on processing information in retinal images, sensations (of all modalities), and the like.[53] On Gibson's view, as a result of evolutionary and survival requirements, sentient animals are capable, at the level of molar functioning, of extracting ("picking up") invariant information embedded in ambient light and other media accessible to their senses, relating to their moving effectively through their respective ecological niches. Curiously, Gibson all but eliminates the cognitive agent: he concedes, in effect, a preestablished (evolutionary) congruity between the perceptual capacities of different species and their ecological space. The information to be extracted is already in the ambient environment; and the different species simply tune in on (spontaneously perceive) information pertinent to their particular mode of survival. The theory is designed to outflank the dichotomies of the subjective and objective – "separate realms of consciousness and matter, . . . psychophysical dualism".[54]

For Gibson, perceptual information is genuinely *real*, actualy *in* the environing media, "*picked up*" without the need for a mediating cognition or mental states. The information is remarkably congenial, for it provides "affordances" – the perception of "a value-rich ecological object": "The *affordances* of the environment [Gibson says] are what it *offers* the animal, what it provides or *furnishes* either for good or ill."[55] An affordance is said to be invariant, "objective, real, and physical", though it "cuts across the dichotomy of subjective-objective . . . . It is equally a fact of the environment and a fact of behavior. It is both physical and psychical, yet neither. An affordance points both ways, to the environment and to the observer".[56]

Gibson insists that the culturally altered environment of humans "is not a *new* environment – an artificial environment distinct from the natural . . . – but the same environment modified by man . . . . It is . . . a mistake to separate the cultural environment from the natural environment, as if there were a world of mental products distinct from the world of material products".[57] He holds that the cultural is perceptually accessible in precisely the same "realist" sense as the natural. But he offers no evidence for this, and his account obscures the fact that much of the import of particular features of the environment *cannot* possibly be detected as he supposes, actually requires a complex measure of learning and an active interpretation of whatever is perceptually accessible – for example, that certain objects are sacred or profane or may affect one's efficiency and the like. It is no accident, therefore, that Gibson has always tried to accommodate, within the scope of ecological optics, the perceptual discrimination of two-dimensional paintings and drawings.[58] But his theory leads to a paradox: that if they are real objects, if they are representational at all, the objects of two-dimensional paintings are merely "virtual objects", incapable of being actually encountered in ecological space; and if they are merely "virtual," then they can hardly be discriminated without applying some learned interpretative canon. Even if one supposed, with Gibson, that the painter intends to fix information about environmental invariances, there could be no "natural" way to "pick" these up – that is, without being oriented to the contingent conventions of a particular culture.[69]

The issue at stake is somewhat more subtle than may at first appear. Gibson replaces the computational intelligence of animals, internalized either in genetic terms or in terms of contingent learning or both, with an evolutionary symbiosis between creature and environment that does not concede internal representations. So seen, his cognitivist opponent could easily claim (as, in effect, Chomsky does)[60] that the difference between fully cognitive and merely cognitive-like internal processing is not important. But, first, the restoration of a "cognizing" agent of some sort dramatizes the difficulty of construing the historically contingent regularities of a culture as no more than unusual coded versions of independently specifiable species-specific ecological invariances. Secondly, by parity of reasoning, precisely because the Gibsonian model, is, however surprisingly, an extreme transform of the cognitivist's own view, we must see that Gibson's difficulties are matched by cognitivist analogues – as in the interpretation of two-dimensional paintings, say, or understanding a natural language. On the other hand, although it is much too thin and too primitive a theory as it stands, Gibson's informational realism (the thesis that, relative to an evolved species, affordances are simply embedded in environing niches) *does* provide a coherent clue regarding explanatory models of intelligence that *are not, or are not exclusively, infra-psychological in nature*. In this sense, the very simplicity of his account demonstrates that the familiar claim – notably championed by Fodor[61] – that there is no conceptual alternative to the computational model, is simply false. In Gibson's system, the causal regularities that explain apparent intelligence or the subcognitive viability of species must be formulated symbiotically, not infra-biologically. It is as if ecological niches were relatively hardwired, not genetic codes. The liberalized theory that we seek, therefore, would favor: (a) a functional system that cannot be satisfactorily explained in terms of any genetic or infra-psychological competence; and (b) a system in which the intelligent behavior of its members cannot be satisfactorily explained in terms of invariances assigned to that emergent system, or to interactions between such a system and its variable environment. The most important point to consider, here, is simply the coherence and plausibility of such a theory. That, after all, is precisely what, intuitively, we construe a *culture* to be.

In this connection, it is not unhelpful to notice that Gibson is inevitably inconsistent about the nature of affordances – which, in effect, preclude the full force of admitting social and cultural learning. "If the affordances of a thing are perceived correctly", Gibson says, "we see that it looks like what it *is*. But we must, of course, *learn* to see what things really are . . . . " (He means, by learning, no more than a fortunate, causally induced adjustment in perceptual orientation that permits ecological pickup under non-standard circumstances.) Again, Gibson says: "According to the theory being developed, if information is picked up perception results; if misinformation is picked up misperception results".[62] And yet, in the same context, he also holds that "When Koffka asserted that 'each thing says what it is,' he failed to mention that it may lie. More exactly, a thing may not look like what it is".[63] But here, obviously, Gibson admits the developed role of an interpretive agent; and, should the theory of ecological invariances fail (as in explaining events within human culture), we should be well on our way to an account of the liberalized sort mentioned.

It is a very small step from here to see that the work of the ethologists – principally, Tinbergen,[64] Lorenz,[65] and von Frisch[66] – may be reasonably construed as specifying a symbiosis very much like Gibson's in the respect at issue, except that it specifically postulates a distinctly preestablished *social* harmony, chiefly though not exclusively species-specific. According to Tinbergen's lucid account, instinctual patterns exhibit a characteristic hierarchical complexity which, in effect, coordinates hardwired "consummatory acts" – relatively "simple and . . . stereotyped movements", "an entirely rigid component, the fixed pattern" of species-specific responses to selective stimuli (for example, the characteristic patterns of chasing, biting, and threatening behavior of fighting stickleback responding to appropriate color and behavioral stimuli) – and "a more or less variable component, the *taxis*, the variability of which . . . is entirely dependent on changes in the outer world," adaptive or "purposive . . . higher patterns" of behavior, softwired, accommodating "learning" but always instrumental to the fixed consummatory acts.[67] Apparently, precisely the same kind of "innate releasing mechanism" is involved, say, in the honey bee's nectar-sipping and the male stickleback's fighting response in the presence of another male within its territory. Tinbergen generalizes, remarking that "The strict dependence of an innate reaction on a certain set of sign stimuli leads to the conclusion that there must be a special neurosensory mechanism that releases the reaction and is responsible for its selective susceptibility to such a very special combination of sign stimuli. This mechanism we will call the Innate Releasing Mechanism . . . ".[68] The work of the ethologists, then, characteristically *substitutes*, for an internal cognitive system of representing the structured features of the external world, a causally effective innate response mechanism that is (ultimately) rigidly coordinated in a preestablished (evolutionary) way with selected external stimuli. A "sign", in Tinbergen's idiom, does not designate an internal element of representation or that stimulus in the external world to which an internal representation corresponds; it signifies only that external stimuli selectively *release*, in a causally efficient way, an innately organized response mechanism.[69] It is the reactive sensitivity of the animal, not its analytic powers, that is decisive.

The work of the ethologists is rather severely circumscribed, as far as the analysis of fully cultural life is concerned. There can be no question about that. Nevertheless, they have confirmed certain very instructive patterns even among largely instinctual species. For one thing, instinctual patterns often appear to be nest- or colony- or hive-wide – that is, gauged to plural, medium-sized natural social units in which strikingly specialized individual members cofunction in some genetically preestablished social way. Even if there were a reason to suppose that the social insects, say, computed their responses to socially provided or socially pertinent stimuli, there would normally be no reason to suppose (parsimoniously) that individual insects (workers, soldiers, and the like) had internalized the organizational purposes of the entire functioning society: their genetically formed competence is still: (a) restricted to their own individual mode of instinctual functioning, though also (b) coordinated by virtue of the *genetic organization of the entire population* so as to serve a preestablished social function. In this way, even in the context of the most hardwired animal behavior – in a context, also, in which the difference between cognitivist and subcognitivist or noncognitivist idioms may seem somewhat verbal – there is a distinct analogue of Saussure's recognition of the unlikelihood or impossibility of internalizing within each member of a functioning society a full representation of that society's interpretive and behavioral rules. It is impossible at the level of human language, and it is otiose at the level of differentiated instinctual societies. The preestablished social harmony of cognitive powers need not be cognitivist in nature; it may for instance be merely chemically triggered, hormonally evolved as in sexual specialization.

Consider some examples once again. Hölldobler remarks that "in the carpenter ants (*Camponotus herculeanus*) it has been demonstrated that the nuptial flights of both sexes are synchronized by a strongly smelling secretion released from the mandibular glands of the males. The males release this synchronizing pheromone during the peak of the swarming activity, at which time the females are stimulated to take off too".[70] But communication is not restricted to chemical means alone; in honey bees and wasps, for instance, acoustical signals and behavioral posturing and ritual (the *Tanzsprache*, most famously) may initiate the releasing mechanism.[71] Among the social wasps, trophallaxis (in effect, regurgitation of food and its exchange or distribution usually among the members of a colony) is often said to be initiated "by a series of tactile signals."[72] The point at issue is simply that the idiom of speaking of "signals" is intended benignly, as in Tinbergen's sense. Otherwise, if it were supposed to signify a more complex form of communication, involving, say, the internal representation of the initiating stimulus and its cognitive retrieval by a particular insect, including the representation of the functional relationship itself, the theory would appear (on all the evidence) both otiose and utterly unlikely. This marks, for example, the conceptual difficulty of subscribing to Donald Griffin's so-called cognitive ethology: "one can [he says] interpret the application of the *Schwanzelntanz* [the waggle dance] to cavities [that is, to the location of sites appropriate for establishing a new colony] as a precisely hardwired, genetically programmed behavior pattern, never used for many generations, but ready, in latent form, to be elicited at the time of swarming. Suppose we allow ourselves to consider the alternative interpretation that foraging bees know what they are doing".[73] In fact, Griffin presses the use of "symbol" and "intention" in the context of bee communication.[74] But, for one thing, the waggle dance need not be construed simply as hardwired. On Tinbergen's model, it can include the variability of the *taxis*. For a second, either the idiom of cognition is a mere *façon de parler* or, given the intended analogy with the human paradigm, its use requires much stronger empirical evidence than has yet been supplied. Thus, Gibson's use of the term "information" is almost deliberately equivocal – signifying either the cognitive content of perceptual pickup *or* the causal efficacy of an ecological system involving organism and environment, in which perceptual pickup (including "learning" and "error") is treated as cognitive only in a heuristic sense. But thirdly, even on a cognitivist reading, there still remains a need to account for the complex cofunctioning of a society, in which the genetics of populations cannot be satisfactorily analyzed in cognitivist terms – or in which the cognizing organism must be taken to be the entire social system rather than any individual member of the aggregated set of organisms.[75]

Here, there is a clear gap in the argument and a need for close reflection on the respects in which the functional regularities of a human culture can both build upon and distinctly direct the mental life and behavior of the individual members of a society, without relying exclusively on genetic – or cognitivist – processes. The ethologists show how social organization can be understood in ways that cannot be cognitivistically ordered (even if they presuppose cognitivist powers), though they tend to rely largely on genetically determined competences. Nevertheless, even here, as in certain of the studies of Konrad Lorenz, the patterns of animal imprinting and the emergence of relatively rigidly transmitted "traditions" do appear; that is, processes "of increasing the *selectivity* of a response elicited through innate releasing mechanisms and directed towards conspecifics, by additional acquisition of complex conditioned Gestalts" (within a particular society – stickleback, greylag geese, jackdaws, pack-dogs).[76] In human cultures, we must assume (if the fully emergent nature of language-based social behavior is to be vindicated) that such selective patterns can be generated and sustained by processes that are relatively independent of (though compatible with) genetic determinants – *a fortiori*, independent of genetic cognitivism (Chomsky's and Fodor's theories, for instance), which need not be denied *some* function – or, where we admit them, cognitivist processes must represent at least the partially internalized powers of contingent, culturally induced regularities. On the hypothesis, here associated with Saussure's thesis, these regularities – preeminently, linguistic – cannot be completely internalized by any process of learning or individual acquisition and cannot be completely in place innately or biologically (*contra* Chomsky).

III

At this point, we need to reflect on various conceptual possibilities bearing on the explanation of human intelligence – in order to appreciate how the alternative causal thesis sketched points the way to an even greater attenuation of the cognitivist thesis, while at the same time it facilitates a plausible admission of the distinctly cognitive character of the cultural life of man. This will enable us to complete this survey.

In general, the phenomena of intelligence and cognition must be construed as psychologically real – although the relevant *predicates* may well be validly extended (as to computers) without psychological import. Since intelligence and cognition are psychologically real, however, their modelling cannot be merely abstractly or functionally justified in an empirical way, without attention to (what may be called) the constraints of incarnation.[77] At the present moment, it is certainly not clear that an AI or cognitivist modelling of the incarnate sort is or is not possible. Our concern here is simply that it is not the only viable or plausible way of proceeding to account for the processes of human intelligence. All accounts that are biologically motivated are committed to *some* range of genetic hardwiring. Among the ethologists, instinctual behavior is effectively construed as combining the organizing forms of a species-specific genetic code (that is finite and fixed) and causal influences permitting the development of variable habits, "dialects", forms of "learning" or imprinting and the like, congruent with such a code and suitably invariant within a given colony or pack or hive. Here, a psychologically rich characterization is all but obviated, though it is not incompatible with applications of a purely instinctual model. In effect, Chomsky has extended the application of a related account, except that, since language constitutes the most complex achievement of natural organisms and underlies the characteristic forms of human reason and freedom, the manifestations of language cannot be construed as merely instinctual. Chomsky must be right in affirming innate capacities underlying linguistic performance. But whether there is a form of linguistic competence that is itself innate, in the sense in which the elements of a universal grammar are in some way genetically coded, is a conceptual as well as an empirical mystery. It is quite possible, for instance, that the hardwired components underlying what Chomsky calls linguistic competence are entirely *sublinguistic* (a prospect more in keeping with the developing picture of the biology of animal communication).[78] In any case, Chomsky's theory *requires the learning of historically formed natural languages that* (though constrained somehow by an innate grammar) *cannot be described merely in terms of the variable taxis of a form of instinctual activity*. Hence, on any plausible view of human language, we need to have an account of the cognitive aspects of the essentially contingent features of a diachronically developing language, more complex than any instinctual model could afford. *A fortiori*, the same will be true of linguistically informed intelligence. *If*, that is, the syntactic structure of a human language *cannot* be adequately accounted for in terms of an innatist theory of grammar – and if other coordinate abilities requisite for speech cannot be similarly explained – even if *some* innate competence (even some proto-linguistic competence) must be admitted, then, assuming that human culture cannot be explained solely in terms of an instinctual model such as that favored by the ethologists, there *must* be a system of cognitive but noncognitivist processes that accounts for the distinctive features of contingent, plural, coherent, and diachronically changing cultures. The following model remark (Chomsky's) must, therefore, be utterly misdirected:

Taking a grammar to be a system of rules that provides representations of sound and meaning (among others), their specific character to be determined as research progresses, our task is to discover the representations that appear and the rules operating on them and relating them: and more important, to discover the system of universal grammar that provides the basis on which they develop. One may think of the genotype as a function that maps a course of experience into the phenotype. In these terms, universal grammar is an element of the genotype that maps a course of experience into a particular grammar that constitutes the system of mature knowledge of a language, a relatively steady state achieved at a certain point in normal life.[79]

The platonist, that is, the nativist with respect to concepts (notably Fodor), offers no prospect of being able to detail a set of innate concepts adequate for the generation of all culturally acquired concepts. The nativist with respect to grammar (notably Chomsky) cannot provide a sufficiently rich and convincingly universal innate grammar that could relevantly minimize the causal importance of a cognitively acquired contingent culture and historical language. The analytic structuralist (Lévi-Strauss, for instance) cannot be certain that the putative rules or laws of cultural organization are even biologically or psychologically real – let alone genuinely comprehensive, "totalized", and empirically accurate.[80] Saussure's and Lévi-Strauss's versions of structuralism are motivated in remarkably similar ways, insofar as they are both distinctly opposed to the "chaos" of historical contingency and doubtful about the psychological reality of the structuring rules they invoke. For example, in a justly famous passage, Lévi-Strauss affirms:

Methodological analysis has not, and cannot have, as its aim to show how men think . . . it is doubtful, to say the least, whether the natives of central Brazil [for example], over and above the fact that they are fascinated by mythological stories, have any understanding of the systems of interrelations to which we reduce them. . . . I therefore claim to show, not how men think in myths, but how myths operate in men's minds without their being aware of the fact. What I am concerned to clarify is not so much what there is in myths (without, incidentally, being in man's consciousness) as the system of axioms and postulates defining the best possible code, capable of conferring a common significance on unconscious formulations which are the work of minds, societies, and civilizations chosen from among those most remote from each other. . . . This is why it would not be wrong to consider this book itself as a myth: it is, as it were, the myth of mythology.[81]

The vagaries of structuralism, however, need not detain us.

We have been searching for clues regarding a viable model for explaining human intelligence, that is not primarily or exclusively cognitivist in nature and not exclusively or predominantly nativist. These lines surely point to at least the following requirements: (i) biologically innate mechanisms that are not cognitively qualified but are causally involved in the generation of cognitively qualified processes; (ii) homuncular or sub-molar cognitive processes that are not innate and are identifiable only as empirically assigned sub-processes of molar cognitive processes; (iii) cognitively qualified or "cognigenic" causal forces that are never completely internalized psychologically at the molar level, or are only partially internalized or internalized only differentially among the members of a smoothly functioning society, and contingently affect, nevertheless, the cognitively qualified or cognitively pregnant ("cognignomic") work and behavior of particular members; and (iv) causal influences that are irreducibly qualified in linguistic ways, linked within a system of language only by a measure of idealizing (though, *per* Saussure, even the psychologically real processes manifested in *parole* are not specifiable independently of idealized *languages*), ascribed as a system only to a determinate society, and subject as a system to diachronic changes of an Intentional sort. Qualifications of these sorts may well lead, among the cultural sciences, to radical adjustments regarding the nature of causality, causal explanation, even the methodology of science itself. Notice that (i)–(iv) progress from the genetic to the cultural, from the noncognitive to the cognitive, from the sub-molar to the molar, from the infra-psychological to the societal. They suggest, therefore, a certain comprehensiveness and flexibility at the same time they feature the *sui generis* nature of cultural processes.

In a relatively straightforward sense, the thrust of these remarks may not unfairly be characterized as Wittgensteinian in inspiration. Of course, it is notoriously difficult to fix Wittgenstein's views – even to attribute particular theories to him. But several of the most salient themes of the *Philosophical Investigations* are effectively focussed in the following very well-known remarks:

F. P. Ramsey once emphasized in conversation with me that logic was a 'normative science'. I do not know exactly what he had in mind, but it was doubtless closely related to what only dawned on me later: namely, that in philosophy we often *compare* the use of words with games and calculi which have fixed rules, but cannot say that someone who is using language *must* be playing such a game. – But if you say that our languages only *approximate* to such calculi you are standing on the very brink of a misunderstanding. For then it may look as if what we were talking about were an *ideal* language. As if our logic were, so to speak, a logic for a vacuum. – Whereas logic does not treat of language – or of thought – in the sense in which a natural science treats of a natural phenomenon, and the most that can be said is that we *construct* ideal languages. But here the word "ideal" is liable to mislead, for it sounds as if these languages were better, more perfect, than our everyday language; and as if it took the logician to shew people at least what a correct sentence looked like.[82]

If language is to be a means of communication there must be agreement not only in definitions but also (queer as this may sound) in judgments. This seems to abolish logic, but does not do so. – It is one thing to describe methods of measurement, and another to obtain and state results of measurement. But what we call "measuring" is partly determined by a certain constancy in results in measurement.[83]

Here, in effect, Wittgenstein corrects the structuralist views of Saussure and the nativist views of Chomsky. The essential themes, which will be at once recognized to be quintessentially Wittgenstein's and which can now be appreciated for their extraordinary subtlety and force, include the following: (a) that knowing how to use a natural language presupposes and entails a speaker's sharing in a strongly convergent way what, in the society that uses that language, are taken to be salient true beliefs; and (b) that knowing how to use a natural language is to participate in an ongoing social practice in which one's own improvisational behavior is consensually tolerated as an instance and as an *extension* of that practice, without reference to antecedently fixed rules or without having been generated by the use of such rules.

Several famous paradoxes are involved here. First, understanding linguistic concepts presupposes agreement with regard to what, precisely, the concepts are first supposed to enable us to grasp;[84] second, social practices and the rules that putatively individuate them and fix what counts as conforming with them are not conceptually independent of one another, are not hierarchically ordered as if in the manner of an axiomatized system, and cannot have their proper scope antecedently fixed – either jointly or one by the other. The idea is that would-be rules are only idealizations of the actual living practices of a society, and are themselves confirmed as such only by reference to the diachronic practices they pretend to organize; also, that particular agents become culturally habituated in accord with the historically contingent practices of a society in such a way that their own behavior remains tolerably congruent with the prevailing patterns of agreement of that society – so that their contribution actually extends the sense of such agreement, to which others (acting later) similarly conform and add. The important thing is that *such social systems are effective in spite of the fact that particular agents (most significantly, infants initiated into the "forms of life" of a particular culture) – though also ourselves – cannot have psychologically internalized their conceptual features completely or finally.* Synchronically, the very existence of linguistic communication (*a fortiori*, the very existence of a culture, of a linguistically informed community) presupposes sufficient species-specific invariants of a biological and psychological sort that would make that phenomenon possible; diachronically, the successful initiation of pre-cultural members of a species into the cultural practices of that very species entails effective influences congruent with such practices functioning below, and incipiently at, the level of a cognitive understanding of those practices themselves. Infants learning a language are influenced (somehow) to conform with the behavioral idiosyncrasies of a culture (probably in a way that habituates what will becomes linguistically relevant but is at first only superfluous, redundant, or congruent with regard to biologically innate or partially innate forms of infra-species communication).[85] Ironically, one may – in the spirit of Chomsky's own sense of rationalist speculation, though contrary to his own doctrine – claim that the human species must be genetically endowed with a *general cognitive competence* to internalize, in a partial but strategic way congruent with the improvisational requirements of cultural rules, the particular rules of all natural languages and natural cultures. There is no greater mystery in this than there is in any nativist account of acquiring a natural language; but there is of course a mystery.

The most important feature of this broadly Wittgensteinian model is its economy and coherence. Obviously, there are a great many detailed questions that its adoption would require to be answered. Two considerations seem particularly helpful to mention – before bringing our discussion to a close. For one, once we construe one's culture as a kind of intensionally organized second nature – incarnate in, and distinctively habituating, one's animal nature – we cannot but be struck by the parallel between instinctual behavior and behavior of choice, between the limited but genuine variability of instinct and *taxis* and the extraordinary variability of cultural practice and personal idiosyncrasy. It seems hardly improbable that if hardwired instincts can accommodate "improvisational" *taxes* under different environmental conditions, softwired cultural regularities should be able to accommodate even freer and more radically improvisational novelties on the part of those who have mastered such practices. Lorenz illustrates the first very neatly:

> What a jackdaw eats, where it seeks its food, the enemies which evoke alarm and flight, and even the nest-sites which are preferred – all are extensively dependent upon the individual's personal experience and in fact upon the "tradition" of the society. With respect to these behavior patterns, we can identify relatively great variability and adaptability. In Northern Russia and Siberia, the jackdaw exhibits no fear whatsoever towards man, nests in any low-built peasant house, constructs the nest mainly with straw and lives on insects trapped on the open ground. In our own large towns, the jackdaw is

extremely timid, nests only on high and inaccessible parts of buildings, constructs its nest with a great variety of materials (particularly employing a lot of paper) and specializes according to its locality on plundering pigeon nests, feeding on refuse, etc. However, the behavior of the birds *with respect to one another* exhibits not the slightest variability. The motor display patterns and vocalizations – together with the corresponding innate responses which ensure the social co-ordination of the colony members – are "photographically" identical.[86]

It is difficult to construe such variability in Griffin's way, "as a precisely hardwired, genetically programmed behavior pattern, never used for many generations, but ready, in latent form, to be elicited at [a suitable] time"; and it is equally difficult to construe it, by analogy with Chomsky's view of language, as the result to the jackdaw's genetic program mapping "a course of experience into the phenotype". The contingent, inventive, local, and still further alterable nature of such mixtures of *taxes* and learning is, however puzzling, quite remarkable. Colonies of the same species vary significantly, and individuals within colonies, differ in ways that cannot be assimilated merely to variations on a common species template, although such variations do not affect hardwired consummatory instincts.

In the human case, it is regularity in the absence of actual fixity of culturally local practices that tempts us to treat the idealized rules of such practices in terms of genetic coding. But cultural regularities are Intentional, which, as far as we can tell, are *sui generis* and irreducible to mere biological uniformities of any sort. Hence, we may speculate, playfully, that, once we concede the fact of human language and its peculiar complexities – in particular, that syntax and semantics are inextricably interdependent and that understanding a natural language cannot be separated from sharing the contingent nonlinguistic experience of a particular society – the very stability of language and culture imitates as much as possible the biological uniformities of our first nature (in which, after all, they are incarnate). Here, the principal novelty rests with regularizing the Intentional itself. Since, on the hypothesis, it cannot be hardwired and since to treat it as softwired is merely to idealize sampled uniformities within an open continuum of diachronic change, the very notion of linguistic and cultural regularity cannot fail to incorporate the Intentional improvisation of apt human agents – whose own behavior affects in greater or lesser ways the fit of alternative idealizations. The culturally softwired is not a program legibly ascribed in a direct way to genetic hardwiring.

One rather densely worked but suggestive formulation of the essential feature of cultural rules and uniformities is afforded in Pierre Bourdieu's conception of the *habitus* – which manages to combine the Wittgensteinian theme of knowing how to "go on" and the Marxist theme (as in the *Theses on Feuerbach* and the *Grundrisse*) of the power of social *praxis* to generate actual practices that evolve in a coherent but inventive way:

The habitus [he says], the durably installed generative principle of regulated improvisations, produces practices that tend to reproduce the regularities immanent in the objective conditions of the production of their generative principle, while adjusting to the demands inscribed as objective potentialities in the situation, as defined by the cognitive and motivating structures making up the habitus.[87]

The *habitus* is "history turned into nature",[88] "an acquired system of generative schemes objectively adjusted to the particular conditions in which it is constituted".[89] Here, Bourdieu recognizes the Intentional distinction of historical uniformities, the impossibility that they be ahistorically projected by structuralist methods (*à la* Lévi-Strauss or Chomsky or the "structuralist readers of Marx"), and the need to provide a culturally relevant causal framework in order to avoid arbitrariness in specifying and ordering those historical uniformities themselves.

The second consideration intended, regarding the Wittgensteinian model, is the inescapability of an interpretive leap – as far as public practices are brought to bear on the behavior and activity of individual and aggregated members of a society. The Intentional import of what anyone does or makes is not fixed or indubitable at any moment. Even in direct speech encounters, such formulations as Grice's account of speakers' intentions suggest the complexity of the most elementary cultural exchanges. Add to this the increasing historical "distance" of the written word, the artifact, the remembered act, and the problematic of the entire hermeneutic tradition stands before us. The vagaries of that movement need not concern us here.[90] But what we must concede, in the context of our earlier reflections, is that human agents cannot be said to know how they have computed their own intelligent utterances; that their intentions are characteristically ephemeral, ambiguous, uncertain and even – most importantly – self-assigned by a later interpretive effort; and that they, as well as others in their society, characteristically identify what was actually uttered and done by means of an interpretive consensus informed by constantly monitoring the diachronically changing coherence of relevant social practices. What was done remains Intentionally significant; but its import is relatively and variably fixed by reference to the same evolving idealizations by which the uniformities of cultural practices are first recognized. This is why, say, the peculiarly interiorized dignity attributed to Rembrandt's mature portraits is (justifiably) assigned on the strength of an interpretation of the movement of an entire era *retrospectively perceived*. There is no alternative. But to say so is to acknowledge a factor that both complicates the methodology of the human sciences and lightens the conceptual puzzle that the nativists have pressed. The complication concerns the holism of human intelligence and the corresponding "intertextuality" of human utterances and artifacts. The one addresses the intentional linkage, within a model of rationality, of distinct psychological states.[91] The other addresses the intensional linkage, within a cultural space, of the meaning of particular texts (actions, utterances, artifacts) and of whatever else a society diachronically collects, remembers, and understands in the distinctive ways it does. This supplies the meaning of that extraordinarily cryptic but accurate pronouncement of Barth's: "Every text, being itself the intertext of another text, belongs to the intertextual, which must not be confused with a text's origins . . . ".[92]

In the context of the foregoing discussion, we may gloss Lorenz's, Bourdieu's, and Barth's remarks as conceding: (i) that, as gifted animals, human beings undoubtedly possess innate competences; (ii) that, as culturally adept persons, human beings undoubtedly internalize a developing grasp of changing social practices; (iii) that, as a highly socialized species, man's cognitive powers are probably facilitated and harmonized by genetically governed processes of a noncognitive sort; (iv) that, as culturally emergent beings, humans are subject to Intentional regularities, that cannot be governed but only constrained by biologically hardwired rules or laws; (v) that, as culturally responsive agents, human beings cannot master the Intentional regularities of the societies in which they are groomed without improvising, relative to a changing environment, utterances and forms of work and behavior that count at once as instantiating and extending the regularities of those very societies; and (vi) that, as the aggregated members of a functioning society, sharing a language and culture that each understands in only a partial, specialized, and contingently limited way, the import of all significant work and behavior, all utterance and artifact, requires a measure of interpretive consensus focused in terms of relatively idealized accounts of the systematic practices of the entire society. No one of these findings, nor their collection, leads us to cognitivism. On the contrary, taken together, they draw us on to a picture of the human sciences radically unlike that of the physical and biological sciences and (at least as they have been recently practiced) of linguistics and the informational sciences.

NOTES

[1] See John Haugeland, 'The Nature and Plausibility of Cognitivism', *The Behavioral and Brain Sciences*, II (1978).
[2] See Daniel Dennett, *Content and Consciousness* (London: Routledge and Kegan Paul, 1969); *Brainstorms* (Montgomery, Vt.: Bradford Books, 1978).
[3] Jerry A. Fodor, *The Language of Thought* (New York: Thomas Y. Crowell, 1975).
[4] Michael Polanyi, *Personal Knowledge*, corr. (Chicago: University of Chicago Press, 1962).
[5] Maurice Merleau-Ponty, *Phenomenology of Perception*, Colin Smith (transl.) (London: Routledge and Kegan Paul, 1962).
[6] See Massimo Piattelli-Palmarini (ed.), *Language and Learning: The Debate between Jean Piaget and Noam Chomsky* (Cambridge: Harvard University Press, 1980).
[7] Noam Chomsky, *Rules and Representations* (New York: Columbia University Press, 1980); Jean Piaget, *Structuralism*, trans. Chininah Maschler (New York: Basic Books, 1970).
[8] A full discussion of these views and of the cognitivist program is provided in Joseph Margolis, *Philosophy of Psychology* (Englewood Cliffs: Prentice-Hall, 1984).
[9] Cf. Joseph Margolis, 'The Trouble with Homuncular Theories', *Philosophy of Science*, XLVII (1980).
[10] H. P. Grice, 'Utterer's Meaning and Intentions', *Philosophical Review*, LXXVIII (1969), 151. Cf. also, 'Meaning', *Philosophical Review*, LXVII (1957); and 'Utterer's Meaning, Sentence-Meaning, and Word-Meaning', *Foundations of Language*, IV (1968).
[11] This explains, perhaps, Daniel Dennett's attraction to Grice's thesis; cf. 'Conditions of Personhood', in Amélie Oksenberg Rorty (ed.), *The Identities of Persons* (Berkeley: University of California Press, 1976).
[12] 'Utterer's Meaning, Sentence-Meaning, and Word-Meaning'.
[13] Grice characteristically shifts from the inten*t*ional to the inten*s*ional, though he never explains or justifies the turn. Whether the change is benign enough is itself a matter of argument; cf. for example, James Cornman, 'Intentionality and Intensionality', *Philosophical Quarterly*, XII (1962).
[14] Cf. Joseph Margolis, 'Meaning, Speakers' Intentions, and Speech Acts,' *Review of Metaphysics*, XXVI (1973); and 7, below; also, Hilary Putnam, 'The Meaning of "Meaning"', in *Philosophical Papers*, Vol. 2 (Cambridge: Cambridge University Press, 1975); *Meaning and the Moral Sciences* (London: Routledge and Kegan Paul, 1978).
[15] Louis Hjelmslev, *Prolegomena to a Theory of Language*, rev. trans. Francis J. Whitfield (Madison: University of Wisconsin Press, 1961), p. 9.
[16] Ferdinand de Saussure, *Course in General Linguistics*, Wade Baskin (transl.), Charles Bally, Albert Sechehaye, Albert Riedlinger (eds.) (New York: Philosophical Library, 1959), pp. 13–14. Page references are to the McGraw-Hill paperback edition, 1966.
[17] Noam Chomsky, *Language and Responsibility*, John Viertel (transl.) (New York: Pantheon Books, 1979), pp. 140, 152–153.
[18] Hjelmslev, *op. cit.*, 14.
[19] *Ibid.*, p. 14–15; cf. pp. 39–40.
[20] Cf. *ibid.*, pp. 21–23.
[21] Hilary Putnam, 'The Nature of Mental States', in *Philosophical Papers*, Vol. 2.

[22] *Loc. cit.*
[23] Cf. *ibid.*, p. 5–6. Contrast Fodor, *loc. cit.*, and Dennett, *loc. cit.*
[24] Alfred Tarski, 'The Semantic Conception of Truth', *Philosophy and Phenomenological Research*, IV (1944). Donald Davidson has, more recently, attempted to sketch a theory in which Tarski's account could be directly applied to natural languages, though the venture seems essentially programmatic; cf. for instance, 'In Defense of Convention T', in Hugues Leblanc (ed.), *Truth, Syntax and Modality* (Amsterdam: North Holland, 1975); and 'Truth and Meaning', *Synthese*, XVII (1967). For a favorable and unfavorable appraisal of Davidson's account, see Gareth Evans and John McDowell (eds.), *Truth and Meaning* (Oxford: Clarendon, 1976); and Ian Hacking, *Why Does Language Matter to Philosophy*? (Cambridge: Cambridge University Press, 1975).
[25] Michael A. Arbib and David Caplan, 'Neurolinguistics Must be Computational', *The Behavioral and Brain Sciences*, II (1979). Their commentators take them to task for this, though one at least (D. Langendoen) argues that "linguistics must be computational too" (470): Cf. 'One Author's Response' (Arbib), *ibid*.
[26] *Loc. cit.*
[27] Cf. Margolis, *Philosophy of Psychology*.
[28] Saussure, *op. cit.*, pp. 81, 84.
[29] *Ibid.*, p. 84.
[30] *Ibid.*, p. 85.
[31] *Loc. cit.*
[32] This general view of Saussure corresponds to the extremely perceptive account that Roland Barthes affords, in *Elements of Semiology*, Annette Lavers and Colin Smith (transl.) (New York: Hill and Wang, 1967), particularly pp. 15–22. (Page references are to the paperback edition.)
[33] Saussure, *op. cit.*, p. 14.
[34] *Ibid.*, p. 110.
[35] *Loc. cit.*
[36] Pierre Bourdieu, *Outline of a Theory of Practice*, trans. Richard Nice (Cambridge: Cambridge University Press, 1977).
[37] *Loc. cit.*
[38] Thomas S. Kuhn, *The Structure of Scientific Revolutions*, 2nd ed. enl. (Chicago: University of Chicago Press, 1970).
[39] Possibly the most extreme – curiously similar – opposition to the possibility of a rational method appears in the work of Paul Feyerabend and Jacques Derrida. See, for instance, Paul Feyerabend, *Against Method* (London: NLB, 1975), and Jacques Derrida, *Speech and Phenomena*, trans. David B. Allison (Evanston: Northwestern University Press, 1973). Cf. Joseph Margolis, 'Wissenschaftliche Methoden und Feyerabends Plädoyer für den Anarchismus', in *Versuchungen, Aufsätze zur Philosophie Paul Feyerabends*, heraus. Hans Peter Duerr (Frankfurt am Main: Suhrkamp, 1981).
[40] Noam Chomsky, *Language and Mind*, enl. ed. (New York: Harcourt Brace Jovanovich, 1972). p. 83.
[41] Cf. Jonathan Culler, *Sturcturalist Poetics* (Ithaca: Cornell University Press, 1975), Ch. 10. Culler considers chiefly the work of Roland Barthes, Jacques Derrida, and Julia Kristeva, in which – sometimes equivocally, the denial of a generative and interpretive system and the proposal of *a système décentré* are conflated.

[42] For example, Chomsky remarks, contrasting his own view with that of Foucault, that "He [that is, Foucault] is, I believe, skeptical about the possibility or the legitimacy of an attempt to place important sources of human knowledge within the human mind, conceived in an ahistorical manner," *Language and Responsibility*, p. 75.

[43] *Language and Mind*, p. 71.

[44] Cf. *Language and Responsibility*, p. 128; also, Justin Leiber, *Noam Chomsky* (New York: St. Martin's Press, 1975), pp. 64–68; and Ned Block, 'Troubles with Functionalism', in C. Wade Savage (ed.), *Minnesota Studies in the Philosophy of Science*, Vol. IX (Minneapolis: University of Minnesota Press, 1978).

[45] Cf. *Language and Mind*, pp. 180–185.

[46] *Loc. cit.*

[47] For an illuminating discussion of the issue, see D. W. Hamlyn, *Experience and the Growth of Understanding* (London: Routledge and Kegan Paul, 1978), Chapter 7–8.

[48] Cf. Noam Chomsky, *Syntactic Structures* (The Hague: Mouton, 1959), Ch. 2.

[49] *Language and Responsibility*, pp. 148–154. Cf. Leiber, *op. cit.* p. 122.

[50] *Ibid.*, p. 193.

[51] J. J. Gibson, *The Ecological Approach to Visual Perception* (Boston: Houghton Mifflin, 1979); and *The Senses Considered as Perceptual Systems* (Boston: Houghton Mifflin, 1966).

[52] *The Ecological Approach to Visual Perception*, p. 2.

[53] Cf. Ulric Neisser, *Cognition and Reality* (San Francisco: W. H. Freeman, 1976).

[54] *The Ecological Approach to Visual Perception*, p. 141.

[55] *Ibid.*, pp. 140, 127. The concept of "affordance" is Gibson's adjusted interpretation of the Gestaltist notion, *Aufforderungscharakter* (sometimes translated "valence"). Cf. p. 138.

[56] *Ibid.*, p. 129.

[57] *Ibid.*, p. 130.

[58] Cf. *ibid.*, Chapter 15; also, J. J. Gibson, 'The Information Available in Pictures,' *Leonardo*, IV (1971); and 'On the Concept of Formless Invariants in Visual Perception', *Leonardo*, VI (1973).

[59] Thus Gibson, *ibid.*: "What modern painters are trying to do, if they only knew it, is paint invariants" (p. 284); "But the essence of a picture is just that its information is not explicit" (p. 285). Cf. Nelson Goodman, *Languages of Art* (Indianapolis: Hackett Publishing Co., 1968).

[60] Chomsky, *Rules and Representations*, pp. 69–70.

[61] Fodor, *The Language of Thought*; cf. also, 'On the Impossibility of Acquiring "More Powerful" Structures', in Massimo Piattelli-Palmarini (ed.), *Language and Learning: the Debate Between Jean Piaget and Noam Chomsky* (Cambridge: Harvard University Press, 1980).

[62] *Op. cit.*, p. 142.

[63] *Ibid.*, p. 143.

[64] N. Tinbergen, *The Study of Instinct* (New York: Oxford University Press, 1951, 1969).

[65] Konrad Lorenz, *Studies in Animal and Human Behavior*, 2 vols., trans. Robert Martin (Cambridge: Harvard University Press, 1970, 1971).

[66] Karl von Frisch, *The Dance Language and Orientation of Bees*, trans. Leigh E. Chadwick (Cambridge: Harvard University Press, 1967).

[67] Tinbergen, *op. cit.*, p. 110.
[68] *Ibid.*, pp. 41–42.
[69] *Ibid.*, pp. 25–37.
[70] Bert Hölldobler, 'Communication in Social Hymenoptera,' in Thomas A. Sebeok (ed.), *How Animals Communicate* (Bloomington: Indiana University Press, 1977), p. 418.
[71] Cf. Karl von Frisch, *loc. cit.*; J. L. Gould, 'Honey Bee Communication: Misdirection of Recruits by Foragers with Covered Ocelli,' *Nature*, CCLII (1974); J. L. Gould, 'Honey Bee Recruitment: The Dance-Language Controversy', *Science*, CLXXXIX (1975).
[72] Hölldobler, *op. cit.*, p. 449; Hölldobler mentions particularly, here, the work of H. Montagner, *Comportements trophallactiques chez les guêpes sociales* (Paris: Service du Film de Recherche Scientifique, Film no. B2053).
[73] Donald R. Griffin, 'Prospects for a Cognitive Ethology', *Behavioral and Brain Sciences*, IV (1978), 535. Cf. Donald R. Griffin, *The Question of Animal Awareness* (New York: Rockefeller University Press, 1976).
[74] *Loc. cit.*
[75] There is some justification for so speaking, at the level of sponges, the Portuguese man-o-war, social amoebae, and the like – even of termites and honey bees. But the fact that the genetics of populations is not straightforwardly reducible to the genetics of aggregated individuals does not entail that populations constitute real biological organisms of any sort. Cf. Theodosius Dobzhansky, *Genetics of the Evolutionary Process* (New York: Columbia University Press, 1970); *Mankind Evolving* (New Haven: Yale University Press, 1962).
[76] 'Part and Parcel in Animal and Human Societies', *op. cit.*, Vol. 2, p. 147; cf. also, 'A Consideration of Methods of Identification of Species-specific Instinctive Behavior Patterns in Birds', *op. cit.*, Vol. 1.
[77] See Margolis, *Philosophy of Psychology*. Cf. Pylyshyn, 'Computational Models and Empirical Constraints', *Behavioral and Brain Sciences*, I (1978); also, the 'Open Peer Commentary' and 'Author's Response'; Noam Chomsky, *Reflections on Language* (New York: Pantheon, 1975); and Hubert L. Dreyfus, *What Computers Can't Do*, rev. ed. (New York: Harper and Row, 1979), especially the introduction to the revised edition.
[78] Cf. for instance, J. D. Miller, 'Speech Perception by the Chinchilla: Voiced-voiceless Distinction in Alveolar Plosive Consonants', *Science*, CXC (1975); and A. M. Liberman *et al.*, 'Perception of the Speech Code', *Psychological Review*, LXXIV (1967).
[79] Noam Chomsky, *Rules and Representations*, p. 65.
[80] Cf. Bourdieu, *op. cit.*: and Lucien Goldmann, *Essays on Method in the Sociology of Literature*, William Q. Boelhower (transl.) (St. Louis: Telos Press, 1980).
[81] Claude Lévi-Strauss, *The Raw and the Cooked*, trans. John and Doreen Weightman (New York: Harper and Row, 1969), p. 12.
[82] Ludwig Wittgenstein, *Philosophical Investigations*, G. E. M. Anscombe (transl.) (New York: Macmillan, 1953), § 81.
[83] *Ibid.*, § 242.
[84] Cf. Hamlyn, *op. cit.*, Chapter 6.
[85] Cf. L. S. Vygotsky, *Thought and Language*, trans. Eugenia Hanfmann and Gertrude Vakar (Cambridge: MIT Press, 1962).
[86] Lorenz, 'Part and Parcel in Animal and Human Species', pp. 147–148.

[87] Bourdieu, *op. cit.*, p. 78.
[88] *Loc. cit.*
[89] *Ibid.*, p. 95.
[90] See particularly Hans-Georg Gadamer, *Truth and Method*, trans. (from 2nd ed.), Garrett Barden and John Cumming (New York: Seabury Press, 1975); and Paul Ricoeur, *Interpretation Theory: Discourse and the Surplus of Meaning* (Fort Worth: Texas Christian University Press, 1977). The latest remnant of (naive) romantic hermeneutics is represented by E. D. Hirsch, Jr., *Validity in Interpretation* (New Haven: Yale University Press, 1967); on Hirsch, cf. Joseph Margolis, *Art and Philosophy* (Atlantic Highlands: Humanities Press, 1980).
[91] Cf. Donald Davidson, 'Mental Events', in Lawrence Foster and J. W. Swanson (eds.), *Experience and Theory* (Amherst: University of Massachusetts Press, 1970).
[92] Roland Barthes, 'From Work to Text', trans. Josué V. Harari, in Josué V. Harari (ed), *Textual Strategies; Perspectives in Post-Structuralist Criticism* (Ithaca: Cornell University Press, 1979), p. 77.

7

WITTGENSTEIN AND NATURAL LANGUAGES: AN ALTERNATIVE TO RATIONALIST AND EMPIRICIST THEORIES

No one denies that language is a cultural phenomenon of some sort, that the fledgeling members of new generations of human beings eventually exhibit their speech – the actual power of *parole* – by some process of natural acquisition, by merely living in an adult community that already shares a language and a history of common experience and work focused and informed by that language. Instructively, the very passage with which Ludwig Wittgenstein begins his *Investigations* (which is largely addressed to that intriguing issue), the passage from St. Augustine's *Confessions*, admits the point in such a way that, being ubiquitous, one is almost inclined to ignore it: "When they (my elders) named some object, and accordingly moved towards something, I saw this and I grasped that the thing was called by the sound they uttered when they meant to point it out."[1]

The passage is remarkably sensible and tantalizingly brief. But it already contains the germ of a promising theory of human culture and language acquisition. Wittgenstein himself chooses to construe Augustine's remarks as favoring "a particular picture of the essence of human language" – roughly, that the individual words of a language are names and that their meanings are the objects for which the words stand.[2] In speaking thus, Wittgenstein deflects our attention from the cultural thesis he appears to share with Augustine, the one that gives his *Bemerkungen* the peculiar, almost unperceived, power that they possess. (He moves on, of course, to an implicit criticism of Augustine.) We may in fact characterize the *Investigations* as a series of rather anecdotal probings of what may fairly be called "cultural space", offered without explicit notice of that intention. Being inexplicit, readers of Wittgenstein, particularly Anglo-American philosophers, have tended not to appreciate its importance. Certainly, the dominant theories of language acquisition and linguistic competence almost eliminate the cultural as a factor complicating the analysis of the nature of language.

Here and there, Wittgenstein tends to be more explicit about cultural and historical variability – for example, in his lectures on aesthetics, where he directly considers differences in "cultured taste", that is, the developed taste of people of different cultures and subcultures and of aspects of historically shifting traditions.[3] But the very way in which he mentions and explores these matters shows that they did not surface for him as a fully determinate topic among others. It would be wrong to say that he ignored the cultural, for that is precisely what all the post-Tractarian material is all about. In effect, he offers an impressionistically deployed series of perceptions about the complicating import of the cultural dimension of human existence – *not* a series of impressions – that brings his philosophical efforts into noticeable accord with the thrust of the more explicit investigations about cultural life favored by the phenomenologists and the hermeneutic philosophers.[4] One might almost construe the division between the *Tractatus* and his later work as Wittgenstein's extraordinary effort to plumb, in a manner peculiarly appropriate to each, the conceptual features of what Dilthey sorted as *Naturwissenschaften* and *Geisteswissenschaften*. So construed, the telling feature of Wittgenstein's work is simply that he appears not to have favored any tendency toward what might be called methodological dualism. In this, he seems even closer in spirit to more contemporary Continental thinkers, though he does not actually discuss – say, in the manner of Heidegger – the historicity of human culture and existence. This is not to say, for instance, that Wittgenstein is directly concerned with epistemological issues in the earlier *Tractatus*, though, even there, he *is* in a sense concerned with its foundations, its possibility, in a way that may be more sanguinely construed than is usual.[5] For example, his interest in mysticism (notoriously ignored in the more serious discussions) is otherwise hard to account for.

Wittgenstein seems to have been particularly focused on what is involved in boarding and sharing a culture, particularly a language, both synchronically and in those diachronic respects in which one finally understands an alien language or succeeds in mastering a native one. This is perhaps why there is a kind of timelessness about his discussion of the linguistically informed behavior of imagined societies. But the use of such terms as "language games" and "forms of life" telescopes the undercurrent themes of historical change and cultural variation that cannot be ignored in any interpretation of the *Investigations*. The point is worth pressing in spite of – perhaps because of – the confused reception of the way in which Peter Winch has applied these Wittgensteinian themes.[6]

There is already, in the *Investigations* and elsewhere, an essential thesis that serves to correct the (largely mistaken) impression of Winch's hermeneutical use of Wittgenstein's views – that bears directly on the theory of language acquisition and competence, and that brings Wittgenstein's contribution into close play with the principal theories about the analysis of language. In the lectures on aesthetics (1938), for instance, Wittgenstein remarks: "If you came to a foreign tribe, whose language you didn't know at all and you wished to know what words corresponded to 'good', 'fine', etc., what would you look for? You would look for smiles, gestures, food, toys. ([Reply to objection:] If you went to Mars and men were spheres with sticks coming out, you wouldn't know what to look for . . .) We don't start from certain words, but from certain occasions or activities."[7] In the same context, he adds: "If you ask yourself how a child learns 'beautiful', 'fine', etc., you find it learns them roughly as interjections. . . . One thing that is immensely important in teaching is exaggerated gestures and facial expressions. The word is taught as a substitute for a facial expression or a gesture." Actually, the point is equally central to the quotation from Augustine, for Augustine goes on to say, speaking of learning language among his elders: "Their intention was shown by their bodily movements, as it were [*tamquam*] the natural language of all peoples: the expression of the face, the play of the eyes, the movement of other parts of the body, and the tone of voice which expresses our state of mind in seeking, having, rejecting, or avoiding something."[8] Here, the key Wittgensteinian themes are these: (i) the learning of a natural language, that is, the natural learning of a language, involves membership in a living society of competent adults, a species-specific human nature, and in particular a ramified range of natural bodily expressions of a prelinguistic sort sharing which permits and facilitates the rapid learning of a language; (ii) learning a language entails the prelinguistic coordination of perception, bodily expression, and forms of behavior that favor certain classificatory or descriptive distinctions rather than others. In effect, Wittgenstein acknowledges that understanding the culturally specific regularities of even an alien society – and an infant's acquisition of the regularities of its own – presupposes and depends upon species-specific forms of natural expressiveness. This is not sufficient, of course, but it cannot fail to offset certain extreme (utterly unnecessary and conceptually impossible) conditions that Winch, for one, is alleged to have proposed or defended. Actually, Winch specifically favors what he calls (following Vico perhaps) "certain fundamental notions . . . 'limiting notions' . . . [the] significance [of which] . . . is that they are inescapably involved in the life of all known human societies in a way which gives us a clue where to look, if we are puzzled about the point of an alien system of institutions".[9] These are not merely forms of expressiveness, though they must be related to them; they are centered rather on certain universal concerns – "birth, death, sexual relations" – that apparently set severely circumscribed limits to the variable behavior of the most diverse human societies. Winch explicitly opposes extreme relativism, therefore,[10] and explicitly admits cross-cultural understanding.

These themes will be seen to be remarkably close to those in current dispute among so-called empiricist and rationalist theories of language acquisition. For example, they suggest biological but non-linguistic universals that are bound to affect the acquisition of a natural language (pertinent to rationalist claims); and they suggest ramified conditions under which the field study of alien languages cannot be taken merely to have imposed our own local conceptual schemes on fathoming the behavior of independent peoples (pertinent to empiricist claims). These labels need not be too misleading if we appreciate that they are intended primarily to identify the specific (empiricist) theories of linguistic mastery proposed by W. V. Quine[11] and B. F. Skinner,[12] on the one hand, and Noam Chomsky,[13] on the other. Chomsky himself is largely responsible for the details of the intended contrast, consequently also for the distortion of his interpretation of Wittgenstein as an empiricist. He addresses himself primarily to the Wittgenstein of the *Blue and Brown Books,*[14] possibly of the *Investigations* as well. But Wittgenstein's remarks elsewhere, as in the lectures on aesthetics and in *Zettel,*[15] show that he holds to a relatively unified picture of language in the post-Tractarian texts. Certainly his is not an empiricist account in the Humean sense – as Chomsky rather more accurately characterizes Quine's theory.[16] What is obviously misleading is the superficial similarity between Wittgenstein's and Quine's attention to the behavior of the natives of an alien society, as far as the field linguist's attempt to grasp their language is concerned. Still, there are enormous differences. First of all, Wittgenstein wished to disabuse philosophers of the habit of looking for a specifically inner mental event or state that provided the (shall we say, Augustinian) meaning of any term of psychological description; but he never encouraged a physicalist reduction of the psychological itself. Secondly, when Wittgenstein spoke about human behavior, he never meant to speak of responses to stimuli in any sense in which either could be characterized or identified adequately in nonintentional terms. Chomsky himself has pressed the point, with compelling force, against both Quine and Skinner.[17] Unfortunately, sensing that Wittgenstein is not committed to the undertaking he himself favors as rationalist or nativist, Chomsky is obliged to classify him as an empiricist. This alone suggests a sense in which both empiricist and rationalist theories of language are inadequate and in need of correction along the very lines that Wittgenstein explored.

In the opening paragraphs of the *Investigations*, when Wittgenstein speaks of behavior, he speaks rather more pointedly of human actions. In the shopping illustration, for instance, in which one takes the slip marked "five red apples" to the shopkeeper, who then acts to filll the order – Wittgenstein's remark is, "*er handelt*", that is, busies himself with things in a certain way[18] – we see at once the sense in which Wittgenstein could not possibly have been a radical behaviorist. The sense in which we attend to another's overt expression and behavior is the key, for Wittgenstein, in which the intentional life of a society (more in the hermeneutic than the psychological sense) may be grasped in whatever idiosyncrasy it exhibits because, ultimately, it is a variant – historically contingent – development of and beyond the natural, species-specific expressiveness of the human face and body. In this regard Wittgenstein is in entire agreement with Augustine. The upshot is that the Wittgensteinian theory of language acquisition entails as well that: (iii) learning a language exploits a tacit responsiveness that permits one to master a practice without being able, or needing, to supply – at the moment of mastery or subsequently – its proper analysis or explanation;[19] and (iv) understanding a language requires familiarity with the peculiar range of experience and work characteristic of this society rather than of that, in the context of which the meaning we assign to words depends on their recognized uses in the life of that society.

Wittgenstein wishes to emphasize that we could not possibly learn a language unless we shared a common human nature in which meaning and intentions are linked by our specific responsiveness and sensitivity to needs, behavior, perception, and desire. So he says, as already noted, words substitute for natural expressions and gestures. But he also says, "We don't understand Chinese gestures any more than Chinese sentences",[20] by which we must understand that both nonverbal and verbal behavior are culturally shaped in such a way that only an extended familiarity with Chinese practices could help us grasp how the form of Chinese communication has developed from a common human heritage. It is in this sense that Wittgenstein could not be a *radical* behaviorist. First of all, as opposed to Quine, Wittgenstein does not hold that what we learn, in learning a language, is a set of sentences associated with certain stimulus-response regularities. He stresses rather that we begin with gestures and activities, not words. Secondly, again unlike Quine, he denies that we could understand an alien language to any extent, without some familiarity with the enveloping cultural practices that are bound to vary from one society to another – in effect, he denies that we could do so in the absence of what Quine calls "analytical hypotheses". This is not to agree with Quine that we cannot understand an alien language except to the extent that we impose *our* "analytical hypotheses" on *their* otherwise largely unintelligible behavior. On Wittgenstein's view, the species-specific expressiveness of the human race (and, by Winch's gloss, the universal human concerns that underlie certain extensions of such expressiveness – for instance, sexual courtship) is so extensive that: (a) the natural acquisition even of mental terms is relatively easy, and (b) the need for applying "analytical hypotheses" in the interpretation of alien (and even of one's own) language and culture is extremely restricted and supplementary at best. On Quine's view, empirical inductions (without "analytical hypotheses") are artificially restricted to no more than "sentences" construed as such by way of a strictly behaviorist reading of "stimulus meaning"; and the imposition of "analytical hypotheses" – of which there may be many alternative systems, incompatible with one another but allegedly compatible with the totality of observed behavior – simply extends the range of what can be assimilated to the essentially parochial conceptual systems of the field observer. As Quine puts it:

> ... only by ... outright projection of prior linguistic habits [analytical hypotheses, can] the linguist ... find general terms in the native language at all, or, having found them, match them with his own; stimulus meanings never suffice to determine even what words are terms, if any, much less what terms are coextensive. The method of analytical hypotheses is a way of catapulting oneself into the jungle language by the momentum of the home language. ... Whatever the details of its expository devices of word translation and syntactical paradigm, the linguist's finished jungle-to-English manual has as its net yield an infinite *semantic correlation* of sentences: the implicit specification of an English sentence, or various roughly interchangeable English sentences, for every one of the infinitely many possible jungle sentences. Most of the semantic correlation is supported only by analytical hypotheses, in their extension between the zone where independent evidence for translation is possible. That those unverifiable translations proceed without mishap must not be taken as pragmatic evidence of good lexicography, for mishap is impossible.[21]

There is a telltale blunder of a linguistic sort in Quine's theory: there can be no individuation of sentences without some incipient specification of constitutive words or terms; but to have admitted that would have entailed a much richer concession regarding the empirically pertinent circumstances under which we examine the behavior of alien peoples than Quine would have been willing to allow. Much that he has to say about the indeterminacy of translation and the impossibility of translational "mishap" depends on this gratuitous assumption. By conceding the social nature of one's own native language, Quine's characteristic insistence that infra- and inter-linguistic communication exhibit the same methodological difficulties is completely undermined; and by acknowledging the universal, species-specific, and social conditions under which all human communication develops, both linguistic and nonlinguistic, we are led to see that analytical hypotheses (in Quine's sense) cannot possibly be fundamental to linguistic comprehension. Nevertheless, as Wittgenstein very clearly observes, the culturally alien remains – only, now, it must be accounted for and resolved in an entirely different way. Wittgenstein explicitly says: "one human being can be a complete enigma to another. We learn this when we come into a strange country with entirely strange traditions; and, what is more, even given a mastery of the country's language. We do not *understand* the people. (And not because of not knowing what they are saying to themselves.) We cannot find our feet with them [*Wir können uns nicht in sie finden*]".[22] There is no way to understand the remark except in terms of the extreme divergence, despite a common human nature, of alternative forms of cultural life.

Chomsky effectively criticizes Quine's behaviorism, but he wrongly assimilates Wittgenstein's views to Quine's.[23] In particular, what Chomsky shows is that there could, on Quine's view, be no effective empirical probabilities favoring one significant stimulus-response correlation over another. This rightly confirms, for Chomsky, that the learning of a language is not the same as the learning of sentences. Quine fails to accommodate the point satisfactorily; and Chomsky concludes, too quickly, that only an innatist conception of language will do. But Wittgenstein explicitly trades not on uninterpreted correlations of physical stimuli and physical responses but on the conceptual condition that language depends upon the natural expressiveness of the human face and body and the species-specific interests of the race. He recognizes, in addition, the divergence of cultures. Nevertheless, holding to the postulate of a common human nature, on which the viability of every variant culture depends, and conceding that, from the vantage of our own culture, we may understand another's culture, Wittgenstein outflanks the incoherencies usually (but, on the best evidence, unjustifiably) attributed to Winch and separates himself from the behaviorists.

His resistance to reductive tendencies has led many, notably Chihara and Fodor,[24] to construe Wittgenstein as a *logical* behaviorist as opposed to a radical behaviorist. These philosophers mean by this that, for Wittgenstein, it is not true that psychological attributes must be definable entirely in terms of physical observables (radical behaviorism), but that it is true that psychological attributes must be conceptually linked with behavioral attributes. In itself, this is not troublesome. But the usual interpretation of logical behaviorism ignores the *cultural* dimension of language and linguistically informed behavior. Characteristically, it results in a muddle about Wittgenstein's concept of a criterion, which is neither a necessary nor a sufficient condition for the occurrence of what it is the criterion of, holds in normal contexts, justifies ascriptions or beliefs, but does not capture the sense of logical entailment (though it provides for it).[25] Now, doubtless, Wittgenstein's conception of a criterion is entirely unsystematic and incomplete, and the adequacy of supposing that all discourse proceeds by means of what Wittgenstein calls criteria is probably untenable. But it is not unreasonable to hold that the (logical) informality of a Wittgensteinian criterion reflects the important fact that criteria are marked as such within the living practice of an actual society – not merely in terms of theoretically possible conditions formulated as necessary or sufficient or necessary and sufficient – and corresponds to the equally puzzling (logical) status of what Wittgenstein calls the rules of a society.

Chihara and Fodor are at some pains to show that ordinary scientific inductions and inferences to the best explanation must account for installing at least some empirical concepts; they cannot all be got from behavioral considerations or from merely conceptual or criterial connections. True enough. But to oppose what they call logical behaviorism this way is both to miss what is distinctive in Wittgenstein's remarks and to leave unexplained an essential feature of language acquisition itself. For one thing, Wittgenstein presses the criterial thesis most strenuously in speaking of ascriptions of mental states: "An 'inner process' stands in need of outward criteria."[26] Secondly, this "conceptual connection" obtains *pre*-linguistically as well as linguistically, both because the face and the body are naturally expressive in a species-specific way and because words replace gestures. Thirdly, on Wittgenstein's view (as on Augustine's), there must be some conceptual connection between natural expression and its linguistic replacement, else no language could be learned in a natural way (that is, by growing up and participating in the life of a human society). And fourthly, learning a language is learning a practice, without one's needing to have learned thereupon how either to define that practice or to explain the logical conditions on which it depends or can be known to function.

This is why Wittgenstein says; "For just where one says 'But don't you *see* . . . ?' the rule is no use, it is what is explained, not what does the explaining";[27] "To guess the meaning of a rule, to grasp it intuitively, could surely mean nothing but: to guess its *application*. And that can't now mean: to guess the kind of application; the rule for it. Nor does guessing come in here";[28] "I cannot describe how (in general) to employ rules, except by *teaching* you, *training* you to employ rules."[29] Here, Wittgenstein is opposing the habit of analysis that, for the sake of an apparent formal clarity based perhaps on logical canons, distinguishes between first-order and second-order insights even with respect to the most elementary use of natural languages. "Does a child," asks Wittgenstein, "learn the sense of multiplication *before* – or after it learns multiplication?"[30]

In short, the learning of a criterion or a rule or a practice *is* the mastering of a practice within the cultural space of some society's form of life. Following a rule *is* a practice: "To obey a rule, to make a report, to give an order, to play a game of chess, are *customs* (uses, institutions)."[31] One cannot judge the adequacy of the practice by reference to a higher-order rule, because the rule itself is to be explained by reference to the practice: "I give the rules of a game. The other party makes a move, perfectly in accord with the rules, whose possibility I had not foreseen, and which spoils the game, that is, as I had wanted it to be."[32] Learning a practice may proceed solely by example[33] and understanding a rule that governs a practice and mastering the living practice itself (or an order) are not separable achievements.[34]

There is no way to understand these remarks *except against the backdrop that language and the activities it informs are the interpenetrating elements of a particular culture*. "What has to be accepted", Wittgenstein declares, "the given, is – so one could say – *forms of life*".[35] This explains the failure of Chihara and Fodor's criticism of Wittgenstein's logical behaviorism. Their central objection is focused on certain putative difficulties in Wittgenstein's view of the learning of "ordinary language mental predicates" by means of criteria. In fact, they wish to press the possibility that " the correct view of the functioning of ordinary language mental predicates would assimilate applying them to the sorts of processes of theorctical inference operative in scientific psychological explanation." But they miss a number of essential points in Wittgenstein's view, and they fail to account satisfactorily for their own. For one thing, they nowhere explain how to avoid the thesis that inner mental states stand in need of outward criteria (at least initially). For a second, they nowhere explain how the initial learning of a particular language can escape dependence on the natural expressiveness of the human face and body; hence, how it can escape dependence on a criterial connection between language and natural expression. For a third, they fail to explain how the inductive procedures of empirical science are initially learned; or even how inductive procedures can succeed in being less informal, logically, than Wittgenstein's criterial connections. For a fourth, they have apparently forgotten that Wittgenstein himself did not at all deny that the concepts and distinctions of ordinary language could be replaced to advantage: "a reform for particular practical purposes [of the ordinary forms of language], an improvement in our terminology designed to prevent misunderstandings in practice, is perfectly possible".[36] And for a fifth, they fail to appreciate the fact that the effectiveness of a criterion – hence, also, of a rule or practice – depends on the shared beliefs and experience of a society, which confirms its sufficiency in normal cases *but not* by means of explicit logical constraints.[37]

Here, we must recall that stunningly simple and profound observation of Wittgenstein's:

> If language is to be a means of communication there must be agreement not only in definitions but also (queer as this may sound) in judgments. This seems to abolish logic, but does not do so. – It is one thing to describe methods of measurement, and another to obtain and state results of measurement. But what we call "measuring" is partly determined by a certain constancy in results in measurement.[38]

There are at least three powerful interlocking themes to be drawn from this remark. In the first place, Wittgenstein is indicating that the effective structure of a living practice cannot be correctly characterized as a formally hierarchical or canonical system. What for a canonical or axiomatized system would be an incoherence – the independence of judgments from the very rules that make them logically formulable – is simply an impoverished and lifeless approximation of the utterly different mode of rigor of a socially shared practice, that comprehends at once agreement in judgments and in the characterization of the conceptual conditions under which such judgments function. Furthermore, this bifurcation itself is the result of reflecting on what the mastery of a language must entail; it is not a necessary or even a familiar part of anyone's gradually acquiring the requisite skill. Secondly, it signifies the impossibility of sharing or understanding a language – not merely in a general way but in as detailed a way as may be pressed – without sharing or understanding the history of actual *parole* and of the extralinguistic experience of the society whose language is at stake. This may be effectively confirmed by examining the rather well-known, often-cited theory of meaning developed by William Alston. The detour should prove doubly constructive: because it will show, however informally, the unlikelihood (*contra* Chomsky) of separating either syntax and semantics or the linguistic and nonlinguistic in terms of accounting for the mastery of a natural language.

Alston makes the bold move of paraphrasing a line from Keats's *Ode to a Nightingale*: "O, for a draught of vintage! that hath been cool'd a long age in the deep-delvèd earth!" He offers: "Oh, for a drink of wine that has been reduced in temperature over a long period in ground with deep furrows in it." He does not, it must be said, believe that there are no important differences between the terms of the one and of the other line; but he asks whether the differences that obtain "are differences in meaning".[39] His straightforward answer is that they are not.

Synonymy and difference in meaning, however, are not only difficult to determine, it is even difficult to say what the grounds might be on which they may be satisfactorily determined. Alston's theory is instructive because, in essaying a fair and close account *of* the meaning of the terms of Keats's line, we are forced to take notice of certain essential problems. Consider for example that Alston substitutes "ground" for "earth". He himself says, " 'Earth' conjures up all sorts of associations – earth mother, fertility, early qualities in people, the source of our being – that are lacking for 'ground'."[40] Now, Alston is prepared to concede that the terms differ in their "associations"; but this, he thinks, "does not involve a difference in meaning. I cannot", he adds,

> see that in saying "It came from the earth" I am taking responsibility for any conditions over and above those for which I am taking responsibility in saying "It came out of the ground". The fact that two words will normally call up different sorts of associations seems to be a fact over and above anything I am disposed to allow myself to be called to account for.[41]

It must be the case, therefore, that Alston is prepared to distinguish, in dealing with a word in a sentential context, between the meaning of a word and concepts, images, possibly beliefs that are in some nonarbitrary way "associated" with that word.

This raises questions of at least two sorts. First, how can we determine the right demarcation between what a word means and what may be associated with a word having the meaning it does? Secondly, how can we determine whether the putative meaning or associations of a given word are properly ascribed to that word itself or to the word in its sentential context or even to the word in its sentential context in its speech-act context, or even to all of that in the context of reasonably assumed extralinguistic background interests and information?

There are other possibilities, of course. For example, the meaning of "draught" in "draught of vintage" seems to be affected by the phrase in which it appears, which, though not an indissoluble idiom, nevertheless does not quite allow (for purposes of comparison) the substitution of "drink" –

though not because of any obvious shift in surface structure. "Draught of vintage" appears to signify wine of a particular and choice season that is drawn or drawn off; and the meaning assigned to "draught" and to "draught of vintage" appears to be affected also by *nonlinguistic information* regarding the storing of wine in kegs below the earth. Then, too, whatever the comparison may be between "draught" and "drink" in other sentential contexts ("I'll have a draft [of beer or ale]", "I'll have a drink [of whisky or orange juice]" – depending on context), it is not clear what the direct bearing of such comparisons may be on the meaning of "draught" in Keats's line. It is clear that the latter cannot simply be decided by examining the use of the former. Similarly, "cool'd" signifies not only "reduced in temperature" but "reduced sufficiently to be cool" – that is, at a temperature suitable for drinking, probably to slake one's thirst. "Reduced in temperature", even associated with "over a long period" does not entail "being cool". "Drink of wine" not only signifies nothing regarding being drawn off or being of a particular season's vintage (after all, a drink could be mixed from the wine of various seasons); it loses for us the meaning that can reasonably be assigned to other words, that is, "cool'd a long age in the deep-delvèd earth" – *and even* the meaning that could have been assigned (perhaps) to "reduced in temperature over a long period in ground with deep furrows in it" *if* Alston's paraphrase had begun with "a draught of vintage". Because it is the meaning that we may assign to "draught of vintage", *in context*, in the context of the actual sentence in the actual discourse of the poem, in the tradition of English poetry, that gives us the point of the sense of the rest of the line: "cool'd" clearly means cooled, in the sense of having been placed in a cool and cooling place. And then, "deep-delvèd earth" reasonably signifies that the soil is furrowed because it has been worked or tilled by farmers: it is the deep-delvèd earth from which the wine comes, the source of the vineyard, and the locale of relevant human labor. But "ground with deep furrows in it" is simply irrelevant: Alston's line could be taken to mean that the speaker wishes, for some mad reason, to have a glass of wine that may be found in ground of a certain distinctive surface appearance, the temperature of which happens to be measurably lowered over a considerable interval of time. Again, "drink of vintage" and "draught of vintage" do not appear to be suitably standard phrases that can be straightforwardly studied as substitutions in standard sentence frames; and "cool'd", in the context of the line and Keats's poem, must have the force of signifying intentional human agency; whereas "reduced in temperature" clearly lacks such meaning. Only if the contexts of comparison were chosen in a way that reflected *this* aspect of the meaning of "cool'd" (if that were possible) could the substitution strategy vacuously confirm the difference in the two expressions. In short, there is absolutely no reason to suppose that "earth" and "ground" and all the other paired terms have, and can be tested to show that they have, the same meaning, differing only in "associations"; there is in fact no clear reason to think that the assignment of particular meanings to particular words and phrases is anything more than a convenient, even necessary, simplification drawn from an entire segment of meaningful discourse – that cannot itself be separated from its normal extralinguistic setting.

The issue at stake is not that of inexact synonyms but rather of how we are to suppose we could begin to provide a fair synonym for a given word – and how syntax may reasonably be supposed to be fixed independently of uncertainties in this regard. Presumably, in order to decide *what a word means*, we must consider how the word is *normally* used in sentential contexts. The trouble with attempting to fix the "normal" use of "draught" or "vintage", with an eye to explicating its meaning in the context of Keats's poem, is, precisely, that we are unclear *which* run of sentences could possibly be expected to illuminate the sense of the terms *as they are used in the line of the poem*, or whether the mere regularities of certain sentence frames actually bear on the precise use of a word in another sentence. Surely, only if one could determine *the* meaning of a word in certain privileged and standard sentential contexts, if reference to such contexts would settle all questions of word meaning, could Alston's distinction between "meaning" and "association" be maintained. What might be judged an association in one sentential context might well be judged the meaning or part of the meaning of a given word in another sentence.

J. L. Austin has pointed out elsewhere that

> it may justly be urged that, properly speaking, what alone has meaning is a *sentence* . . . [that] the sense in which a word or a phrase "has a meaning" is derivative from the sense in which a sentence "has a meaning": to say a word or a phrase "has a meaning" is to say that there are sentences in which it occurs which "have meanings": and to know the meaning which the word or phrase has, is to know the meanings of sentences in which it occurs. All the dictionary can do when we "look up the meaning of a word" is to suggest aids to the understanding of sentences in which it occurs. Hence it appears correct to say that what "has meaning" in the primary sense is the sentence.[42]

Still, even here, there is a careless turn of phrase; for if the meaning of words is derivative from the meaning of sentences, it is not the case, if particular words (used in sentences) have meanings, that to know the meaning of a word is to know the "meanings of the sentences in which it occurs". In fact, it looks very much as if, if it makes sense to ask for the meaning of a particular word (in the context of a sentence), then it makes sense as well to ask what the meaning-of-a-word is – which, an apparent question distinct from asking what the meaning of a particular words is, Austin took to be a piece of nonsense.[43] Moreover, there is good reason to suppose that the meaning of a sentence is impenetrable if one does not have a fair grasp of the meaning of the words that compose that sentence; the notions of word and sentence are correlative – there are no sentences without words, or words without sentences, regardless of how fragmentary particular utterances may be.[44] It is in fact Alston's intention, explicitly in the spirit of Austin's theory of illocutionary acts, to attempt a general theory of the meaning of a word. Roughly, on Alston's view, the meaning of a word is a function of the "illocutionary-act potential" of the sentences in which it occurs. Or, more precisely,

A meaning of W_1 is W_2 = df. In most sentences in which W_2 occurs, W_1 can be substituted for it without changing the illocutionary-act potential of the sentence.[45]

"W_1" of course signifies a word. Two sentences are said to have the same illocutionary-act potential if they are "commonly used to perform the same illocutionary act"; and two words are said (in a thoroughly Austinian spirit) to have the same illocutionary-act potential if "they make the same contribution to the illocutionary-act potential of the sentences in which they occur".[46] But there appears to be no viable basis on which the substitution strategy can be employed, where, that is, sentence frames are not inert, are sensitive instead to contingent linguistic habit and extralinguistic experience.

So Alston's theory cannot be an adequate theory of the meaning of a word, but it is not a nonsense theory: it is a false theory. Furthermore, where the strategy can be employed, there is, as we have seen, no reason to suppose that it may serve to fix the meaning of a word used in a sentence that is not semantically inert. But it is the conviction that the strategy can so serve that appears to justify the distinction between the meaning of a word and associations and connotations that are not part of the meaning of a word. Monroe Beardsley, who subscribes to Alston's thesis (with some *caveats*) is led to certain extreme views about the objective interpretation of literature: in discussing the meanings of the words of Wordsworth's controversial Lucy poem, for instance, he claims in a telltale way that "there really is something in the poem that we are trying to dig out, though it is elusive," and he appears to adopt the view (which affects interpretation) that "connotations and suggestions are not a part of meaning but something psychological and personal

. . . [in effect] subjective and relative."[47] Both Alston's and Beardsley's views, therefore, suppose that the demarcation between meanings and associations can be made out in principle. It is, on the contrary, Austin's conviction that this is impossible: in terms of the present account, Austin clearly implies that no sentences are semantically inert or that the meaning of a particular word can be given by reference to its behavior merely in terms of those that are.

Interestingly, in this connection, Paul Ziff remarks that "many facets of a word's meaning may not be indicated in the corpus; 'faerie' occurs in Spenser's *Faerie Queene*, but neither that text nor the other recorded uses of the word are sufficient to enable one to say exactly what the word meant".[48] Ziff's reason is simply that no regularities found in actual uses can be counted on to exhaust all nondeviant uses that *could* have been made of the word, *and* that the meaning of a word requires attention to that consideration. Ziff also offers a subtler theory of the meaning of a word than does Alston, and, in doing so, he too implicitly rejects Austin's claim regarding the-meaning-of-a-word. Reflecting on Ziff's theory leads us to notice the extreme positions that any reasonable theory would wish to avoid. If we held that all sentences were semantically potent, not inert, in the sense that the environing words of a sentence in which a particular word occurs always decisively affect the meaning of that word, without necessarily affecting syntax, then either there would be no point to dictionary definitions of words or else their function would be regularly misdescribed; for there would be no regularities among the meanings of words that one could rely on in moving from sentence to sentence. This would appear to be close to Austin's conviction that dictionary entries are really aids to understanding the meaning of *sentences*. But that would mean, roughly, that a word never means the same thing in different (type) sentences, which seems very doubtful. On the other hand, the thesis that all sentences – or at least the semantically decisive ones – are semantically inert or isomorphically potent, in the sense intended, leads to the view that the meaning of a word can pretty well be fixed, allowing for the comprehensiveness that Ziff speaks about, if only we could select certain normal or strategic sentence frames to test for meanings. This appears to be pretty close to Alston's conviction. It entails the view that words are combined to form sentences by an at least implicit use of *semantic rules*.

In fact, Alston says as much: " . . . it is the rules which are constitutive of illocutionary acts that are crucial for meaning. For according to [the] definitions [given earlier], meaning is a function of illocutionary-act potential".[49] It is clear, from Alston's account, that the rules constitutive of illocutionary acts require that, "in addition to the utterance of an appropriate sentence, . . . not that certain environmental conditions actually hold or even that the speaker believe them to hold, but only that he take responsibility for their holding".[50] This is a subtle view designed to accommodate, for instance, insincere promises, impossible commands, superfluous assertions, and the like. It also draws attention to the extremely important fact that the rules of speech – if there are any rules – cannot be restricted to the formation of utterances but must include more than narrowly linguistic considerations, such as intention, presupposition, and expectation. This does suggest that the theory of language is a specialized part of the theory of human action and behavior and that, therefore, the meaning of given words and sentences is a function of larger regularities that obtain in the full context of human existence. Nevertheless, there are at least two serious difficulties with Alston's views: (i) the alleged constitutive rules are really rules (if allowed at all) concerning the position or status of a speaker or agent once he is taken to have performed a particular illocutionary act, not rules governing the production of such acts so as to facilitate detecting semantic uniformities among either words or sentences; and (ii) the alleged rules specify only what is entailed in performing a particular illocutionary act rather than the conditions under which it may be reliably detected to be of this or that sort.

Alston notes that Ziff's theory of meaning (and Quine's, also, on his view) represents a divergent development of the thesis he himself subscribes to, namely, that "the meaning of an expression is a function of the conditions under which it is uttered".[51] But Ziff's thesis is quite radically different, for Ziff holds that "meaning is essentially (though not necessarily simply) a matter of (nonsyntactic) semantic regularities".[52] In fact, he explicitly denies that natural languages exhibit rules, even though, at certain stages of instruction, one is often taught what purport to be the rules of grammar and the like. "I am concerned with regularities", Ziff insists; "I am not concerned with rules. Rules have virtually nothing to do with speaking or understanding a natural language".[53] He offers the example of the lovely and syntactically deviant line from Gerard Manley Hopkins' *The Windhover*:

> My heart in hiding
> Stirred for a bird, – the achieve of, the mastery of the thing!

"The regularities", he says, "found in or in connection with a language are not sources of constraint [as, presumably, genuine rules would be]."[54] Deviant utterances – notably the deviant utterances of poetry – do not violate anything, least of all the rules of syntax and meaning; they merely deviate from whatever regularities language exhibits and, in doing so, impose a burden on our effort to understand the meaning of what is said. Furthermore, Ziff maintains that "semantic regularities are not simply regularities pertaining exclusively to linguistic elements: they include but are not restricted to such regularities. Semantic regularities are regularities of some sort to be found in connection with the corpus pertaining to both linguistic elements and other things, e.g., to utterances and situations, or to phrases and persons, as well as to utterances and utterances."[55] Also, and rightly noted, "the corpus of utterances [of any natural language] is continually changing ... the various factors responsible for the utterance of utterances are continually changing". So, although there are regularities to be found, shifting diachronically, there appear to be no closed rules to be formulated.[56] But clearly, if Ziff holds to this thesis, he *cannot* subscribe to the substitutivity thesis of Alston's. And in fact, he remarks that "the claim that a certain syntactic or semantic regularity is to be found in or in connection with [the corpus of utterances of a natural language] and not merely in or in connection with some proper part of [it] can generally be defeated by uttering an utterance that deviates from this regularity".[57] So the possibility of deviation from semantic regularities is itself a necessary condition of the relevance of such a regularity in the analysis of such a corpus of utterances. It will do no good, in considering the meaning of a word in a particular utterance, then, to try to hold to the alleged rule-governed patterns exhibited in "most sentences in which W_2 occurs"; not only are statistical considerations indecisive but the possibility that a word may be used in a semantically deviant way in a particular utterance is itself entailed by the thesis that a language exhibits semantic regularities, *and* that its deviant and regular patterns are indissolubly linked to the extralinguistic conditions under which language actually functions.

The third theme to be drawn from Wittgenstein's remark is perhaps the most shadowy, though it will seem the most obvious: we cannot understand the judgments of a society without understanding its concepts. How important and how easily misunderstood the claim is may be judged from the extravagance of the standard charges made against Winch's well-known theory and his own unsatisfactorily indirect defense. Sometimes, it is true, Winch gives the impression that certain critical concepts – particularly those like *rationality, reality, magic*, and *science* – can only be understood within the comprehensive culture in which they function as they do; and that, therefore, outsiders cannot possibly understand an alien culture. But nothing that he actually says need bear the weight of this interpretation; and, when confronted with it,[58] Winch explicitly and convincingly eschews it. How vexing the issue may be may be seen at once from the following remark – offered in speaking sympathetically of the religion of the Old Testament:

God's reality is certainly independent of what any man may care to think, but what that reality amounts to can only be seen from the religious tradition in which the concept of God is used, and this use is very unlike the use of scientific concepts, say of theoretical entities. The point is that it is *within* the religious use of language that the conception of God's reality has its place, though, I repeat, this does not mean that it is at the mercy of what anyone cares to say: if this were so, God would have no reality .... Reality is not what gives language sense. What is real and what is unreal shows itself *in* the sense that language has. Further, both the distinction between the real and the unreal and the concept of agreement with reality themselves belong to our language. I will not say that they are concepts of the language like any other, since it is clear that they occupy a commanding, and in a sense a limiting, position there. We can imagine a language with no concept of, say, wetness, but hardly one in which there is no way of distinguishing the real from the unreal. Nevertheless we could not in fact distinguish the real from the unreal without understanding the way this distinction operates in the language.[59]

Clearly, Winch recognizes: (a) that one not native to the language and culture of the Hebrews can understand *their* concept of God and compare their notion of God's reality with his own scientifically centered notion of reality; (b) that the concept *reality* is, in some sense, effectively ineliminable among any languaged people, though to say so is not to deny the possibility of extreme divergences – amounting perhaps to enabling strongly incompatible claims about what is real to be favored by different societies (at least as seen from the presumed universal adequacy of the conceptual scheme of any one); (c) that reality itself cannot, in some sense, be properly subsumed as an artifact of the conceptual contingencies of any society, though what can be marked as real cannot be so marked (as if in accord with some naive version of the correspondence theory – possibly like the one that informs Wittgenstein's own *Tractatus*) independently of the conceptual network that actually organizes human thought constitutively; and (d) that it cannot be taken for granted that the concepts of any one culture (however fundamental or powerful – like those of Western science) can be counted on to match, include, accommodate, improve upon, or unilaterally justify an "objective" appraisal of, the concepts of another culture.

There is, indeed, a conceptual relativism in Winch's thesis – very much like Wittgenstein's (as when Wittgenstein speaks of our failing to understand Chinese sentences and Chinese gestures) – but it is a relativism that neither precludes cross-cultural comprehension nor denies cross-cultural uniformities and universals. Winch's essential claim is this: "the concepts used by primitive peoples can only be interpreted in the context of the way of life of those peoples".[60] This may be construed to mean (very wrongly, as I. C. Jarvie does)[61] that only primitive people can interpret the concepts they themselves use; that cultures are cleanly demarcatable one from another; and that all tend to belong exclusively to one culture or another. Or, it may be construed to mean that a proper cross-cultural study and assessment of the concepts of primitive peoples (or of others) cannot proceed without first understanding how those concepts actually function in their original cultures. Winch had applauded Evans-Pritchard's exposure of the modern Western bias in speaking of the irrationality of the Azande – recognizing the relativized nature of the concept of rationality (without – though this is admittedly not altogether too clear in Winch's own account – precluding a cross-culturally adequate notion of rationality or of logical consistency); but Winch went on to criticize Evans-Pritchard himself for his own rather unthinking (and ultimately inconsistent) characterization of the scientific in terms of what was "in accord with objective reality".[62]

Winch never actually provides a systematic account of the general issue of how to understand the claims of realism and idealism,[63] which lurk behind the issue of cultural relativism. But the fact remains that it is entirely possible to support a form of realism (freed from the correspondence theory of truth, hence also not opposed to an equally attenuated idealism – one, say, that restricts truth claims to manageable processes of verification and confirmation) compatible with the strong requirements of culturally relativized systems of concepts. It is at any rate in this spirit that Winch clearly opposes, on the one hand, what he terms "an extreme Protagorean relativism"[64] and, on the other, states the irresistibly obvious, a token of the minimal concession of cross-cultural studies: "*we* do not initially have a category that looks at all like the Zande category of magic. Since it is we who want to understand the Zande category, it appears that the onus is on us to extend our understanding so as to make room for the Zande category, rather than to insist on seeing it in terms of our own ready-made distinction between science and non-science. Certainly the sort of understanding we seek requires that we see the Zande category in relation to our own ready-made distinction between science and non-science."[65] Apart from methodological problems that the cultural sciences are bound to generate, a grasp of the general Wittgensteinian bent of these remarks clearly indicates the hopelessness of attempting to defend, on empirical grounds, something like a Chomskyan theory of language *without a developed account of the actual functional relationship between the putative genetic determinants of linguistic competence and the culturally relativized aptitudes of the aggregated speakers of particular societies.* We need not claim, here, that the two cannot be reconciled. It is enough to say that Chomsky strongly resists the very pertinence of the undertaking.

This leads to a further, absolutely crucial feature of Wittgenstein's conception of language – "The concept of a living being has the same indeterminacy as that of a language"[66] – namely: (v) language is incapable of being completely formalized, in addition to its not needing to be formalized in order to be understood; and (vi) its informality is due to the fact that the meaning of what is said depends on the determinate context of use and experience shared by the membership of a living culture. Language is not such that, as in the Chomskyan sense, an innate finite grammar may generate the infinitely many sentences of a natural language;[67] or, as in the manner recently advocated by Donald Davidson, a finite axiomatized system of sentences forming a theory of truth for a language conforming with Tarski's so-called Convention T will yield an account of the meanings of the infinitely many sentences of a natural language, in terms of their syntactic structures.[68] The innatism of Chomsky does not touch at all on the relevance of the cultural context in which language appears to be mastered. And Davidson's program, provisionally compatible with, but indifferent to, the cultural variety of natural languages, is noticeably and prematurely committed to the elimination of intensional concepts. In fact, it is reasonably clear that, among others, Davidson views Wittgenstein as an opponent; speaking of a commentator's remarks, he says: "I share his bias in favor of extensional first-order languages; I am glad to keep him company in search for an explicitly semantical theory that recursively accounts for the meanings of sentences in terms of their structures . . . I think treatments of language prosper when they avoid uncritical evocation of the concepts of convention, linguistic rule, linguistic practice, or language games."[69] But the intensional – specifically, what we have termed the Intentional – is peculiarly central to the acknowledgement of a cultural context. Apparently, Chomsky eschews it; and Davidson believes he can eliminate it by reduction.

Put in a somewhat different way, the intentional features of cultural experience, in terms of which linguistic mastery is activated, is, for Chomsky, hardly more than an adjunct of a "reminiscence theory". Chomsky actually introduces the notion – admittedly argumentatively – in order to expose the apriorism of the empiricist theory of language acquisition.[70] It is certainly true that he does not subscribe to a "pure reminiscence theory"; he admits, for instance, that the development of the brain many give some support to the empiricist theory, and he concedes, in principle, the contribution of ordinary experience to lingustic performance. But he never really explains how what is perceptually acquired by the child – organized at the surface of experience in some pertinently complex way – activates, or could activate, putatively innate, unlearned, universally applicable, adequate linguistic rules by means of which the generation of all surface utterances and the surface processing of sentences could be satisfactorily managed. Chomsky's theory threatens to be a platonism of some sort – particularly when he speculates about the innate structure of the semantic dimension of language learning. In fact, its most extreme possibility – a fully platonic doctrine of the reminiscence of concepts (from which all possible concepts are somehow generated) – has recently been defended by Jerry Fodor, largely in sympathy with Chomsky's own theory of language or with Chomsky's theory at a particular stage of development.[71]

There are at least two essential difficulties with Chomsky's account: for one thing, Chomsky noticeably neglects to give an explicit description of a child's initially learning to sort, recognize, and reproduce the linguistic utterances of his elders (Augustine's puzzle) – a matter which Wittgenstein tries to clarify by attention to the socially perceived forms of life of a human culture; for another, he neglects to explain the full sense in which linguistic rules *could* be innate and generatively adequate for all the apparently culturally fashioned sentences of distinct human societies. It is, of course, essential to Chomsky's theory that "the surface structure [of sentences] is often misleading and uninformative and that our knowledge of language involves properties of a much more abstract nature, not indicated directly in the surface structure".[72] But this makes all the more baffling the process by which the culturally contingent features of the surface structure of sentences, characteristically providing "degenerate and restricted data",[73] could possibly be sorted in an operationally effective way at the innate level so that species-specific, culturally indifferent, universal, unchanging, and unlearned linguistic rules could be fitted to those data. That is, innatism requires, in addition to a knowledge of underlying rules, a knowledge of appropriate criteria on the satisfaction of which appropriate rules are to be selectively invoked; knowledge of this second kind cannot possibly be innate. Knowledge of how to *use* the rules must, in Wittgensteinian terms, be knowledge *of* the rules; but, there is no plausible way, avoiding an extreme platonism that Plato himself eschews, to construe the first sort of knowledge as innate. In fact, there is every reason to believe that Chomsky holds the empirical support for an innate linguistic competence to be dependent only on the global convergence of natural languages, *not on the detailed generation of the "degenerate and restricted data" of the parole of actual speakers, from specific innate universals*. This is why he disengages his innatism from the behavioral evidence that, originally, he conceded might well test and even decide the issue.[74]

As a corollary to his innatism, Chomsky adds a peculiar emergentism that he refuses to characterize in general evolutionary terms: language, he says, is so utterly unlike animal communication systems that "it seems rather pointless ... to speculate about the evolution of human language from simpler systems".[75] It is, however, not only the biological continuity of human speech with animal communication systems that demands explanation; it is also the very idea that a system of grammatical rules, "an abstract system underlying behavior, a system constituted by rules that interact to determine the form and intrinsic meaning of a potentially infinite number of sentences ... a generative grammar",[76] *could* be biologically transmitted from generation to generation that must be explained. The idea (once suitably elaborated) appears to defy any familiar form of genetic transmission.

Most recently, Chomsky has explicitly maintained that "linguistic theory (or 'universal grammar') is what we may suppose to be biologically given, a genetically determined property of the species: the child does not learn this theory, but rather applies it in developing knowledge of language".[77] (Noticeably, the form of storage, retrieval, and application is nowhere developed.) In order to secure the doctrine, Chomsky must subscribe to a (psychologically) realist theory of syntax; hence, to the "autonomy of syntax", in particular, its independence of any semantic base.[78] But *if*, as the empiricists already argue – notably, Quine (on Chomsky's admission) – "grammatical concepts must be defined on the basis of semantic notions,"[79] and *if, contra* Jerry Fodor and Jerrold Katz,[80] there probably could not be a portion of the deep structure of language that would provide for "a complete characterization of the semantic properties of all utterances of all languages independent of all extralinguistic considerations",[81] then serious difficulties confront Chomsky's undertaking. Chomsky is skeptical about the prospects of a universal semantics, since it appears that semantic representation cannot be separated from "beliefs and knowledge about the world".[82] But this means that the putatively deep grammatical system that Chomsky postulates *may* actually, in principle, depend on the surface features of speech produced by creatures subject to shifting beliefs and experience. If so, then both innatism and linguistic realism remain to some extent open to fundamental challenge, and the very characterization of the regularities of a natural language would become strongly intensional in formulation, culturally relativized, incomplete, cognitively accessible at the surface level, and at least partially heuristic in nature.

Both these lines of argument are peculiarly inhospitable to the cultural context of language acquisition and performance. Rationalism retreats from the cultural world of actual acquisition, postulating an interior, inaccessible agent (or agents) already provided with the universal rules for characterizing the syntactic (possibly, also, the semantic) properties of all utterances of all languages. Empiricism concedes the initial data of natural languages, together with the entire range of culturally recognized beliefs, experience, and the like; but whatever is most distinctive of culture itself – meanings, rules, practices, institutions, historical traditions, propositional attitudes infected by these – it claims to be able to eliminate or replace by means of a completely nonintensional canon. Although Wittgenstein did not address himself explicitly to the dispute here sketched, his conception of criteria that rest on a tacit knowledge of the practices of a culture (in fact, more "know-how" than knowledge), of rules embedded in practices that cannot be completely formalized, of the inseparability of questions of meaning and questions of belief, of the divergence and open-ended nature of cultural life suggests a view of language utterly at variance with both the rationalist and the empiricist theories.

Correspondingly, Davidson's program is a purely formal one. It concedes the cultural phenomena of language, but holds that the meanings of sentences produced in normal human situations can always, in principle, be accounted for in terms of a theory that treats truth-values extensionally, at least in part by the reduction of intentional idioms and the elimination or extensional "regimentation" of intensional concepts, On any fair view, Davidson's program is a very large promissory note, which Tarski himself would have been unwilling to tender.[83] It is, in fact, quite unreasonable to suppose that the program is made more plausible by Davidson's general arguments in favor of reductionism (what he calls "anomolous monism"), both because its neutrality is presupposed by the reductive venture and because the viability of the reductionism is actually dependent on the viability of the program.[84]

But we have still to locate the full distinction and force of Wittgenstein's conception of language acquisition and mastery. Against the empiricists – if we understand by the term, chiefly, Quine, Skinner, Davidson – Wittgenstein would, in effect, have resisted the reduction of the analysis of intentional phenomena in extensional terms. (This, of course, is a philosophical fantasy;

but, for purposes of argument, it is very nearly impossible not to ascribe to Wittgenstein something like a theory.) Against the rationalists – by which term we should understand Chomsky and Katz and Fodor at least – Wittgenstein would, in effect, have resisted the introduction of subterranean psychological powers that work innately with the very concepts and language that persons at the molar level are conceded to use. The distinction of the behaviorists – notably, Skinner and Quine – is, precisely, that, on their view (for rather different reasons, as it happens), mentalistic or intentional discourse can be replaced in principle by a behavioral or physical idiom that is entirely extensional in nature. It is, of course, not necessary that empiricists restrict themselves to the molar or surface level of human activity. For instance, Daniel Dennett has attempted to defend a theory in which, once again, the cultural, intentional, mentalistic complexities of human discourse and human experience and activity are systematically eliminated in favor of homuncular activity at some molecular level, that, eventually, may be described entirely in nonintentional terms.[85] But Dennett relies largely on Quine's having shown the way to eliminate intensional contexts, which, on independent grounds, cannot be said to have as yet succeeded.[86]

In effect, then, Wittgenstein's alternative to rationalism and empiricism may be construed in the following way. The acquisition of language presupposes the prelinguistic, natural expressiveness of the human face and body – which is species-specific; but a "form of life" is the contingent, culturally shaped complex of practices that a society shares, that flowers in variable ways from community to community. Fledgling members (and even alien observers) of a particular society cannot understand its rules and practices without coming to understand the background of shared beliefs and experience in terms of which the accomplished agents of that society become capable – tacitly, at least in part – of knowing how to apply a concept, a criterion, a rule or the like. *If* the empiricists are right (against the rationalists) – though even the semantically oriented among the rationalists (those who advocate the spirit if not the letter of generative semantics) appear to concede the point – then it is impossible to formulate, for natural languages, an innate extensional syntax or a syntax independent of initial semantic constraints; *and* it is impossible to formulate a semantics independent of the beliefs of particular societies. And *if* the rationalists are right (against the empiricists – though even some empiricists concede the point), then it is impossible to account for the linguistic and cultural achievement of human societies solely in terms of some doctrine of contingent learning or acquisition centered on behavioral responses to occasional stimuli.

Wittgenstein implicitly preserves *both* theses, eschewing both rationalism and empiricism. Chomsky's argument is largely that if "nonlinguistic factors" must be incorporated "into grammar: beliefs, attitudes, etc.", then, though "such a move cannot be ruled out" *a priori*, "if it proves to be correct, I [that is, Chomsky] would conclude that language is a chaos that is not worth studying"; it would entail "a rejection of the initial idealization that language appears to require, that is, some invariant . . . 'universal grammar'."[87] But Chomsky himself is quite candid to admit that "the legitimacy of the idealization to language" may be challenged – "the autonomy of syntax" and related notions – and that, for instance, "there are languages where word order" seems quite radically free, where the ordering of the base component itself may even depend on nonlinguistic rules.[88] But if so, then it seems quite premature, to say the least, that, if languages do not exhibit the full-blown nativism Chomsky has recently returned to favor, they must constitute an utter chaos. Davidson's position, on the other hand, is simply not an argument. It is pointless to pretend that an extensional reduction of intentional contexts and intensional concepts (like rules and practices) can be achieved without actually providing the argument. But if it is not, or cannot be, provided, then we are led, by Wittgenstein's instruction, to think of the theory of language not in terms of an autonomous, species-specific syntax or in terms of an autonomous, globally adequate formal semantics, but rather in terms of the historically contingent, divergent forms of life and action and belief of particular cultural communities. On that view, the rules of language, however apparently convergent toward some idealized regularities, are inextricably bound to the contingencies of experience; and the meanings of human utterances, however approximately adequate (in certain *restricted* contexts) the explicit canons of a thoroughly extensional paraphrase, are determinable only informally within the persistent Intentional life of a particular community. To concede the dual thesis is, then, to follow Wittgenstein in a program of linguistic analysis quite radically different from either of the two principal schools of thought contending at the moment. To have reached this point in the argument is to have made the merest beginning. The larger issues that open promisingly now require a fresh approach. Perhaps, however, it would be useful, in bringing this study to a close, to identify, without further comment, a number of the most strategic theses that bridge what we have so far considered and those ulterior questions that subtend them, that still need to be focused, that have in some measure already shaped our sense of the inquiries just completed, and that obviously could support a rich set of further inquiries of an even more comprehensive sort. These must include the following at least: (i) that language is *sui generis*, emergent, irreducible in physical terms, and real; (ii) that language is the paradigm of the Intentional and the condition *sine qua non* of cultural phenomena; (iii) that all extensionalist programs fitted to human psychology, human history, human science, human culture are inextricably dependent, conceptually, on interests and criteria formulable only by linguistically apt agents who are themselves Intentionally emergent and real; (iv) that linguistic and cultural phenomena are not infra-psychologically or infra-biologically reducible; (v) that Intentional phenomena are embodied phenomena and are not dualistic; (vi) that, although only individual persons and aggregates of persons are the cultural agents of cultural activity, Intentional phenomena are or presuppose the irreducibly social sharing of practices, traditions, rules, principles, ideologies and the like; (vii) that all and only culturally emergent entities and phenomena have histories; (viii) that cultures, construed as networks of practices, have histories as well; (ix) that there are real Intentional causes; (x) that causality does not entail nomologicality; (xi) that causal explanation need not be deductive in principle; (xii) that in the human sciences at least, phenomena may be causally explained under, possibly only under, covering institutions; (xiii) that causal explanation of any kind is historically contingent; (xiv) that no science can escape the Intentional constraints of human existence; (xv) that natural languages and natural cultures cannot be totalized, that is, formulated as alternative sets of infinitely many possible well-formed types (of sentences, actions or the like) generated or even heuristically generable by finite sets of rules collecting such sets as systems; (xvi) that any finite segment of such languages and cultures can be construed actually or heuristically as a system; (xvii) that the entire set of natural languages and cultures cannot be totalized, *a fortiori* cannot be totalized genetically; (xviii) that naturally acquired cultural aptitudes formed within a given society characteristically include aptitudes for understanding, or responding acceptably to, deviant and improvisationally novel utterances, acts and the like, and for generating such utterances and acts; (xix) that the Intentional import of Intentionally generated phenomena is a function of consensually generated interpretations of such phenomena by the apt members of a naturally acquired culture; (xx) that interpretive consensus itself has a history; (xxi) that infra-cultural and trans-cultural understanding presupposes and depends upon biologically grounded uniformities adequate to justify a species-specific model of rationality; (xxii) that the apt members of any natural human culture are capable of learning to function acceptably in any other such culture; (xxiii) that cultures are ineliminable parts of the real world, are constantly made and remade and changed by the activity of culturally emergent agents, themselves altered in the process; (xxiv) that at least some parts of the real world manifest historical change; (xxv) that any plausible reformulation of the unified science conception must accommodate the foregoing theses.

NOTES

[1] Ludwig Wittgenstein, *Philosophical Investigations*, trans. G. E. M. Anscombe (New York: Macmillan, 1953), §1.

[2] *Loc. cit.*

[3] *L. Wittgenstein, Lectures and Conversations on Aesthetics, Psychology and Religious Beliefs*, Cyril Barrett (ed.) (Berkeley: University of California Press, 1967): hereafter, *Lectures*.

[4] See Hans-Georg Gadamer, *Truth and Method*, trans. (from 2nd ed.) Garrett Barden and John Cumming (New York: Seabury Press, 1975); Maurice Merleau-Ponty, *The Primacy of Perception* (trans.), ed. James M. Edie (Evanston: Northwestern University Press, 1964); and Jürgen Habermas, *Knowledge and Human Interests*, trans. Jeremy J. Shapiro (Boston: Beacon Press, 1971).

[5] For instance, *contra* G. E. M. Anscombe, *An Introduction to Wittgenstein's Tractatus* (London: Hutchinson, 1959), Chapters 1, 12.

[6] The principal publications by Winch (in this regard) include: *The Idea of a Social Science* (London: Routledge and Kegan Paul, 1958); 'Understanding a Primitive Society', *American Philosophical Quarterly*, I (1964); 'Comment', in Robert Borger and Frank Coiffi (eds.), *Explanation in the Behavioral Sciences* (Cambridge: Cambridge University Press, 1970). Most of the responses to Winch, while on the whole sensible about conceptual dangers risked by his position, are – it must be said in all candor – generally carried away by their own convictions. A very close and careful reading of Winch's texts, I'm afraid, does not *quite* support the extreme charges brought against him. There is no particular reason to suppose that he was not prone to (or even naive about) certain methodological difficulties in his own view. He certainly never took adequate steps to correct impressions conveyed by his own statements – even in response to objections. But in retrospect his critics do seem to have been more than usually wildeyed. The most salient of the critical papers, including a number that Winch himself responded to, include: Alasdair MacIntyre, 'The Idea of a Social Science', *Proceedings of the Aristotelian Society*, Suppl. Vol. LXI (1967); Alasdair MacIntyre, 'Is Understanding Religion Compatible with Believing?' in John Hick (ed.), *Faith and the Philosophers* (London: Macmillan, 1964); Alasdair MacIntyre, 'A Mistake about Causality in Social Science', in Peter Laslett and W. G. Runciman (eds.), *Philosophy, Politics and Society*, Second Series (Oxford: Basil Blackwell, 1962); I. C. Jarvie, 'Understanding and Explanation in Sociology and Social Anthropology', in Borger and Cioffi, *op. cit.*; and A. R. Louch, *Explanation and Human Action* (Oxford: Basil Blackwill, 1966). Winch does commit himself at least to two views that appear mistaken (which MacIntyre duly notes): that all human action is "rule-governed" and that (having) reasons for one's behavior cannot be accommodated as such within causal explanation. But these difficulties, important as

they are, do not entail the radical relativism, incoherence, and impossibility of cross-cultural understanding usually attributed to Winch's putative position.

[7] *Lectures*, pp. 2–3.

[8] *Philosophical Investigations*, §1.

[9] 'Understanding a Primitive Society'.

[10] Peter Winch, 'Nature and Convention', *Proceedings of the Aristotelian Society*, Vol. LX (1959–60).

[11] W. V. Quine, *Word and Object* (Cambridge: MIT Press, 1960); *The Roots of Reference* (LaSalle, Ill.: Open Court, 1974).

[12] B. F. Skinner, *Verbal Behavior* (New York: Appleton-Century-Crofts, 1957).

[13] Noam Chomsky, *Cartesian Linguistics* (New York: Harper and Row, 1966); *Language and Mind*, rev. (New York: Harcourt, Brace, 1972).

[14] Noam Chomsky, 'Some Empirical Assumptions in Modern Philosophy of Language', in Sidney Morgenbesser *et al.* (eds.), *Philosophy, Science and Method: Essays in Honor of Ernest Nagel* (New York: St. Martin's Press, 1969), hereafter, 'Empirical Assumptions'.

[15] Ludwig Wittgenstein, *Zettel*, G. E. M. Anscombe and G. H. von Wright (eds.), G. E. M. Anscombe, transl. (Berkeley: University of California Press, 1970).

[16] 'Empirical Assumptions'.

[17] Noam Chomsky, Review of B. F. Skinner, *Verbal Behavior, Language* XXXV (1959); 'Empirical Assumptions'. Cf. also, Joseph Margolis, *Persons and Minds* (Dordrecht: D. Reidel, 1978), Chapter 11.

[18] *Philosophical Investigations*, §1.

[19] Cf. Maurice Merleau-Ponty, *Phenomenology of Perception* trans. Colin Smith (London: Routledge and Kegal Paul, 1962); and Michael Polanyi, *Personal Knowledge*, corr. (Chicago: University of Chicago Press, 1962).

[20] *Zettel*, §219.

[21] Quine, *op. cit.*, pp. 70–71.

[22] *Philosophical Investigations*, Pt. II, p. 223.

[23] 'Empirical Assumptions'.

[24] Charles S. Chihara and Jerry A. Fodor, 'Operationalism and Ordinary Language: A Critique of Wittgenstein', *American Philosophical Quarterly*, II (1965).

[25] Cf. Rogers Albritton, 'On Wittgenstein's Use of the Term "Criterion" ', *Journal of Philosophy*, LXI (1959), reprinted, with an additional note, in George Pitcher (ed.), *Wittgenstein: The Philosophical Investigations* (Garden City: Anchor Books, 1966).

[26] *Philosophical Investigations*, §580.

[27] *Zettel*, §302.

[28] *Ibid.*, §306.

[29] *Ibid.*, §318.

[30] *Ibid.*, §324.

[31] *Philosophical Investigations*, §199.

[32] *Zettel*, §293.

[33] *Ibid.*, §295.

[34] *Ibid.*, §286.

[35] *Philosophical Investigations*, Pt. II, p. 226.

[36] *Ibid.*, §136.

[37] *Ibid.*, §142.

[38] *Ibid.*, §292.

[39] William P. Alston, *Philosophy of Language* (Englewood Cliffs: Prentice-Hall, 1964), p. 45.
[40] *Loc. cit.*
[41] *Ibid.*, p. 46.
[42] J. L. Austin, 'The Meaning of a Word', in *Philosophical Papers*, J. O. Urmson and G. J. Warnock (eds.) (Oxford: Clarendon, 1961), p. 24.
[43] This redeems to some extent Stuart Hampshire's inquiry about meaning, though not necessarily his theory about meaning; cf. 'Ideas, Propositions and Signs', *Proceedings of the Aristotelian Society*, Vol. XL (1939–40).
[44] Cf. Joseph Margolis, 'Quine on Observationality and Translation', *Foundations of Language*, IV (1968); and 'Behaviorism and Alien Language', *Philosophia*, III (1973).
[45] Alston, *op. cit.*, p. 38. Cf. Joseph Margolis, 'Meaning, Speakers' Intentions, and Speech Acts', *Review of Metaphysics*, XXVI (1973).
[46] *Ibid.*, pp. 36–37.
[47] Monroe C. Beardsley, *The Possibility of Criticism* (Detroit: Wayne State University Press, 1970), pp. 47–48.
[48] Paul Ziff, *Semantic Analysis* (Ithaca: Cornell University Press, 1960), p. 7.
[49] Alston, *op. cit.*, p. 42.
[50] *Ibid.*, p. 42f.
[51] *Ibid.*, p. 108, notes to Chapter 1.
[52] Ziff, *op. cit.*, p. 42. Ziff's intentions about the use of "meaning" are opposed to Austin's: he finds it odd to speak of the meaning of utterances and sentences, not of words, p. 149.
[53] *Ibid.*, p. 34.
[54] *Ibid.*, p. 36.
[55] *Ibid.*, p. 27.
[56] *Ibid.*, p. 22. Cf. also, Hilary Putnam, 'Is Semantics Possible?' and 'The Meaning of "meaning" ', in *Philosophical Papers*, Vol. 2 (Cambridge: Cambridge University Press, 1975). Putnam shows implicitly why, for even "natural kind words," a substitution strategy cannot establish the meanings of words.
[57] *Ibid.*, p. 24; also, p. 57.
[58] Particularly by Jarvie, *loc. cit.*
[59] 'Understanding a Primitive Society'; reprinted in Fred R. Dallmayr and Thomas A. McCarthy (eds.), *Understanding and Social Inquiry* (Notre Dame: University of Notre Dame Press, 1977), p. 162. (Page references are given to the Dallmayr and McCarthy printing, for convenience.)
[60] *Ibid.*, p. 173.
[61] *Loc. cit.*
[62] *Loc. cit.*; cf. E. E. Evans-Pritchard, *Witchcraft, Oracles and Magic Among the Azande* (Oxford: Clarendon, 1937).
[63] Cf. Joseph Margolis, 'Cognitive Issues in the Realist-Idealist Dispute', *Midwest Studies in Philosophy*, V (1980).
[64] Winch, *op. cit.*, p. 161.
[65] *Ibid.*, p. 179.
[66] *Zettel*, §326.
[67] Cf. *Language and Mind.*
[68] Cf. Donald Davidson, 'Truth and Meaning', *Synthese*, XVII (1967); 'In Defense of

Convention T', in Hugues Leblanc (ed.), *Truth, Syntax and Modality* (Amsterdam: North Holland Publishing Co., 1973); and Gareth Evans and John McDowell (eds.), *Truth and Meaning* (Oxford: Clarendon, 1976).

[69] Donald Davidson, 'Reply to Foster', in Evans and McDowell, *op. cit.*, p. 33.

[70] 'Empirical Assumptions'.

[71] Jerry A. Fodor, *The Language of Thought* (New York: Crowell, 1975). Cf. also Jerry A. Fodor, *Representations* (Cambridge: MIT Press, 1981) and 'On the Impossibility of Acquiring "More Powerful" Structures', in Massimo Piattelli-Palmarini (ed.), *Language and Learning: The Debate between Jean Piaget and Noam Chomsky* (Cambridge: Harvard University Press, 1979).

[72] *Language and Mind*, p. 37.

[73] *Ibid.*, p. 64.

[74] Noam Chomsky, *Syntactic Structures* (The Hague: Mouton, 1965), Chapter 2. It is important to notice that Chomsky's suggested "behavioral" criterion (p. 13) is quite capable of accommodating the *caveats* he subsequently mentions in this chapter.

[75] *Language and Mind*, p. 70.

[76] *Ibid.*, p. 71.

[77] Noam Chomsky, *Language and Responsibility* (New York: Pantheon Press, 1977), p. 140.

[78] *Ibid.*, Chapter 6.

[79] *Ibid.*, p. 138.

[80] J. J. Katz, *The Underlying Reality of Language and Its Philosophical Import* (New York: Harper and Row, 1971); *Semantic Theory* (New York: Harper and Row, 1972).

[81] *Language and Responsibility*, p. 141.

[82] *Ibid.*, p. 142.

[83] Alfred Tarski, 'The Concept of Truth in Formalized Languages', *Logic, Semantics, Metamathematics*, J. H. Woodger (transl.) (Oxford: Clarendon, 1956).

[84] Donald Davidson, 'Mental Events', in Lawrence Foster and J. M. Swanson (eds.), *Experience and Theory* (Amherst: University of Massachusetts Press, 1970).

[85] Daniel Dennett, *Content and Consciousness* (London: Routledge and Kegan Paul, 1969); *Brainstorms* (Montgomery, Vt.: Bradford Books, 1978).

[86] Cf. Joseph Margolis, 'The Stubborn Opacity of Belief Contexts', *Theoria*, XLIII (1977).

[87] Chomsky, *Language and Responsibility*, pp. 152–153, 172.

[88] *Ibid.*, pp. 188–189, 193–194.

INDEX

SYNTHESE LIBRARY

Studies in Epistemology, Logic, Methodology,
and Philosophy of Science

1. J. M. Bochénski, *A Precis of Mathematical Logic.* 1959.
2. P. L. Guiraud, *Problèmes et méthodes de la statistique linguistique.* 1960.
3. Hans Freudenthal (ed.), *The Concept and the Role of the Model in Mathematics and Natural and Social Sciences.* 1961.
4. Evert W. Beth, *Formal Methods. An Introduction to Symbolic Logic and the Study of Effective Operations in Arithmetic and Logic.* 1962.
5. B. H. Kazemier and D. Vuysje (eds.), *Logic and Language. Studies Dedicated to Professor Rudolf Carnap on the Occasion of His Seventieth Birthday.* 1962.
6. Marx W. Wartofsky (ed.), *Proceedings of the Boston Colloquium for the Philosophy of Science 1961-1962.* Boston Studies in the Philosophy of Science, Volume I. 1963.
7. A. A. Zinov'ev, *Philosophical Problems of Many-Valued Logic.* 1963.
8. Georges Gurvitch, *The Spectrum of Social Time.* 1964.
9. Paul Lorenzen, *Formal Logic.* 1965.
10. Robert S. Cohen and Marx W. Wartofsky (eds.), *In Honor of Philipp Frank.* Boston Studies in the Philosophy of Science, Volume II. 1965.
11. Evert W. Beth, *Mathematical Thought. An Introduction to the Philosophy of Mathematics.* 1965.
12. Evert W. Beth and Jean Piaget, *Mathematical Epistemology and Psychology.* 1966.
13. Guido Küng, *Ontology and the Logistic Analysis of Language. An Enquiry into the Contemporary Views on Universals.* 1967.
14. Robert S. Cohen and Marx W. Wartofsky (eds.), *Proceedings of the Boston Colloquium for the Philosophy of Science 1964-1966. In Memory of Norwood Russell Hanson.* Boston Studies in the Philosophy of Science, Volume III. 1967.
15. C. D. Broad, *Induction, Probability, and Causation. Selected Papers.* 1968.
16. Günther Patzig, *Aristotle's Theory of the Syllogism. A Logical-Philosophical Study of Book A of the Prior Analytics.* 1968.
17. Nicholas Rescher, *Topics in Philosophical Logic.* 1968.
18. Robert S. Cohen and Marx W. Wartofsky (eds.), *Proceedings of the Boston Colloquium for the Philosophy of Science 1966-1968.* Boston Studies in the Philosophy of Science, Volume IV. 1969.

19. Robert S. Cohen and Marx W. Wartofsky (eds.), *Proceedings of the Boston Colloquium for the Philosophy of Science 1966-1968*. Boston Studies in the Philosophy of Science, Volume V. 1969.
20. J. W. Davis, D. J. Hockney, and W. K. Wilson (eds.), *Philosophical Logic*. 1969.
21. D. Davidson and J. Hintikka (eds.), *Words and Objections. Essays on the Work of W. V. Quine*. 1969.
22. Patrick Suppes, *Studies in the Methodology and Foundations of Science. Selected Papers from 1911 to 1969*. 1969.
23. Jaakko Hintikka, *Models for Modalities. Selected Essays*. 1969.
24. Nicholas Rescher *et al.* (eds.), *Essays in Honor of Carl G. Hempel. A Tribute on the Occasion of His Sixty-Fifth Birthday*. 1969.
25. P. V. Tavanec (ed.), *Problems of the Logic of Scientific Knowledge*. 1969.
26. Marshall Swain (ed.), *Induction, Acceptance, and Rational Belief*. 1970.
27. Robert S. Cohen and Raymond J. Seeger (eds.), *Ernst Mach: Physicist and Philosopher*. Boston Studies in the Philosophy of Science, Volume VI. 1970.
28. Jaakko Hintikka and Patrick Suppes, *Information and Inference*. 1970.
29. Karel Lambert, *Philosophical Problems in Logic. Some Recent Developments*. 1970.
30. Rolf A. Eberle, *Nominalistic Systems*. 1970.
31. Paul Weingartner and Gerhard Zecha (eds.), *Induction, Physics, and Ethics*. 1970.
32. Evert W. Beth, *Aspects of Modern Logic*. 1970.
33. Risto Hilpinen (ed.), *Deontic Logic: Introductory and Systematic Readings*. 1971.
34. Jean-Louis Krivine, *Introduction to Axiomatic Set Theory*. 1971.
35. Joseph D. Sneed, *The Logical Sstructure of Mathematical Physics*. 1971.
36. Carl R. Kordig, *The Justification of Scientific Change*. 1971.
37. Milic Capek, *Bergson and Modern Physics*. Boston Studies in the Philosophy of Science, Volume VII. 1971.
38. Norwood Russell Hanson, *What I Do Not Believe, and Other Essays* (ed. by Stephen Toulmin and Harry Woolf). 1971.
39. Roger C. Buck and Robert S. Cohen (eds.), *PSA 1970. In Memory of Rudolf Carnap*. Boston Studies in the Philosophy of Science, Volume VIII. 1971.
40. Donald Davidson and Gilbert Harman (eds.), *Semantics of Natural Language*. 1972.
41. Yehoshua Bar-Hillel (ed.), *Pragmatics of Natural Languages*. 1971.
42. Sören Stenlund, *Combinators, λ-Terms and Proof Theory*. 1972.
43. Martin Strauss, *Modern Physics and Its Philosophy. Selected Papers in the Logic, History, and Philosophy of Science*. 1972.
44. Mario Bunge, *Method, Model and Matter*. 1973.
45. Mario Bunge, *Philosophy of Physics*. 1973.
46. A. A. Zinov'ev, *Foundations of the Logical Theory of Scientific Knowledge (Complex Logic)*. (Revised and enlarged English edition with an appendix by G. A. Smirnov, E. A. Sidorenka, A. M. Fedina, and L. A. Bobrova.) Boston Studies in the Philosophy of Science, Volume IX. 1973.
47. Ladislav Tondl, *Scientific Procedures*. Boston Studies in the Philosophy of Science, Volume X. 1973.
48. Norwood Russell Hanson, *Constellations and Conjectures* (ed. by Willard C. Humphreys, Jr.). 1973.

49. K. J. J. Hintikka, J. M. E. Moravcsik, and P. Suppes (eds.), *Approaches to Natural Language.* 1973.
50. Mario Bunge (ed.), *Exact Philosophy – Problems, Tools, and Goals.* 1973.
51. Radu J. Bogdan and Ilkka Niiniluoto (eds.), *Logic, Language, and Probability.* 1973.
52. Glenn Pearce and Patrick Maynard (eds.), *Conceptual Change.* 1973.
53. Ilkka Niiniluoto and Raimo Tuomela, *Theoretical Concepts and Hypothetico-Inductive Inference.* 1973.
54. Roland Fraissé, *Course of Mathematical Logic* – Volume 1: *Relation and Logical Formula.* 1973.
55. Adolf Grünbaum, *Philosophical Problems of Space and Time.* (Second, enlarged edition.) Boston Studies in the Philosophy of Science, Volume XII. 1973.
56. Patrick Suppes (ed.), *Space, Time, and Geometry.* 1973.
57. Hans Kelsen, *Essays in Legal and Moral Philosophy* (selected and introduced by Ota Weinberger). 1973.
58. R. J. Seeger and Robert S. Cohen (eds.), *Philosophical Foundations of Science.* Boston Studies in the Philosophy of Science, Volume XI. 1974.
59. Robert S. Cohen and Marx W. Wartofsky (eds.), *Logical and Epistemological Studies in Contemporary Physics.* Boston Studies in the Philosophy of Science, Volume XIII. 1973.
60. Robert S. Cohen and Marx W. Wartofsky (eds.), *Methodological and Historical Essays in the Natural and Social Sciences. Proceedings of the Boston Colloquium for the Philosophy of Science 1969-1972.* Boston Studies in the Philosophy of Science, Volume XIV. 1974.
61. Robert S. Cohen, J. J. Stachel, and Marx W. Wartofsky (eds.), *For Dirk Struik. Scientific, Historical and Political Essays in Honor of Dirk J. Struik.* Boston Studies in the Philosophy of Science, Volume XV. 1974.
62. Kazimierz Ajdukiewicz, *Pragmatic Logic* (transl. from the Polish by Olgierd Wojtasiewicz). 1974.
63. Sören Stenlund (ed.), *Logical Theory and Semantic Analysis. Essays Dedicated to Stig Kanger on His Fiftieth Birthday.* 1974.
64. Kenneth F. Schaffner and Robert S. Cohen (eds.), *Proceedings of the 1972 Biennial Meeting, Philosophy of Science Association.* Boston Studies in the Philosophy of Science, Volume XX. 1974.
65. Henry E. Kyburg, Jr., *The Logical Foundations of Statistical Inference.* 1974.
66. Marjorie Grene, *The Understanding of Nature. Essays in the Philosophy of Biology.* Boston Studies in the Philosophy of Science, Volume XXIII. 1974.
67. Jan M. Broekman, *Structuralism: Moscow, Prague, Paris.* 1974.
68. Norman Geschwind, *Selected Papers on Language and the Brain.* Boston Studies in the Philosophy of Science, Volume XVI. 1974.
69. Roland Fraissé, *Course of Mathematical Logic* – Volume 2: *Model Theory.* 1974.
70. Andrzej Grzegorczyk, *An Outline of Mathematical Logic. Fundamental Results and Notions Explained with All Details.* 1974.
71. Franz von Kutschera, *Philosophy of Language.* 1975.
72. Juha Manninen and Raimo Tuomela (eds.), *Essays on Explanation and Understanding. Studies in the Foundations of Humanities and Social Sciences.* 1976.

73. Jaakko Hintikka (ed.), *Rudolf Carnap, Logical Empiricist. Materials and Perspectives.* 1975.
74. Milic Capek (ed.), *The Concepts of Space and Time. Their Structure and Their Development.* Boston Studies in the Philosophy of Science, Volume XXII. 1976.
75. Jaakko Hintikka and Unto Remes, *The Method of Analysis. Its Geometrical Origin and Its General Significance.* Boston Studies in the Philosophy of Science, Volume XXV. 1974.
76. John Emery Murdoch and Edith Dudley Sylla, *The Cultural Context of Medieval Learning.* Boston Studies in the Philosophy of Science, Volume XXVI. 1975.
77. Stefan Amsterdamski, *Between Experience and Metaphysics. Philosophical Problems of the Evolution of Science.* Boston Studies in the Philosophy of Science, Volume XXXV. 1975.
78. Patrick Suppes (ed.), *Logic and Probability in Quantum Mechanics.* 1976.
79. Hermann von Helmholtz: *Epistemological Writings. The Paul Hertz/Moritz Schlick Centenary Edition of 1921 with Notes and Commentary by the Editors.* (Newly translated by Malcolm F. Lowe. Edited, with an Introduction and Bibliography, by Robert S. Cohen and Yehuda Elkana.) Boston Studies in the Philosophy of Science, Volume XXXVII. 1977.
80. Joseph Agassi, *Science in Flux.* Boston Studies in the Philosophy of Science, Volume XXVIII. 1975.
81. Sandra G. Harding (ed.), *Can Theories Be Refuted? Essays on the Duhem-Quine Thesis.* 1976.
82. Stefan Nowak, *Methodology of Sociological Research. General Problems.* 1977.
83. Jean Piaget, Jean-Blaise Grize, Alina Szeminska, and Vinh Bang, *Epistemology and Psychology of Functions.* 1977.
84. Marjorie Grene and Everett Mendelsohn (eds.), *Topics in the Philosophy of Biology.* Boston Studies in the Philosophy of Science, Volume XXVII. 1976.
85. E. Fischbein, *The Intuitive Sources of Probabilistic Thinking in Children.* 1975.
86. Ernest W. Adams, *The Logic of Conditionals. An Application of Probability to Deductive Logic.* 1975.
87. Marian Przelecki and Ryszard Wójcicki (eds.), *Twenty-Five Years of Logical Methodology in Poland.* 1977.
88. J. Topolski, *The Methodology of History.* 1976.
89. A. Kasher (ed.), *Language in Focus: Foundations, Methods and Systems. Essays Dedicated to Yehoshua Bar-Hillel.* Boston Studies in the Philosophy of Science, Volume XLIII. 1976.
90. Jaakko Hintikka, *The Intentions of Intentionality and Other New Models for Modalities.* 1975.
91. Wolfgang Stegmüller, *Collected Papers on Epistemology, Philosophy of Science and History of Philosophy.* 2 Volumes. 1977.
92. Dov M. Gabbay, *Investigations in Modal and Tense Logics with Applications to Problems in Philosophy and Linguistics.* 1976.
93. Radu J. Bogdan, *Local Induction.* 1976.
94. Stefan Nowak, *Understanding and Prediction. Essays in the Methodology of Social and Behavioral Theories.* 1976.
95. Peter Mittelstaedt, *Philosophical Problems of Modern Physics.* Boston Studies in the Philosophy of Science, Volume XVIII. 1976.

96. Gerald Holton and William Blanpied (eds.), *Science and Its Public: The Changing Relationship*. Boston Studies in the Philosophy of Science, Volume XXXIII. 1976.
97. Myles Brand and Douglas Walton (eds.), *Action Theory*. 1976.
98. Paul Gochet, *Outline of a Nominalist Theory of Proposition. An Essay in the Theory of Meaning*. 1980.
99. R. S. Cohen, P. K. Feyerabend, and M. W. Wartofsky (eds.), *Essays in Memory of Imre Lakatos*. Boston Studies in the Philosophy of Science, Volume XXXIX. 1976.
100. R. S. Cohen and J. J. Stachel (eds.), *Selected Papers of Léon Rosenfeld*. Boston Studies in the Philosophy of Science, Volume XXI. 1978.
101. R. S. Cohen, C. A. Hooker, A. C. Michalos, and J. W. van Evra (eds.), *PSA 1974: Proceedings of the 1974 Biennial Meeting of the Philosophy of Science Association*. Boston Studies in the Philosophy of Science, Volume XXXII. 1976.
102. Yehuda Fried and Joseph Agassi, *Paranoia: A Study in Diagnosis*. Boston Studies in the Philosophy of Science, Volume L. 1976.
103. Marian Przelecki, Klemens Szaniawski, and Ryszard Wójcicki (eds.), *Formal Methods in the Methodology of Empirical Sciences*. 1976.
104. John M. Vickers, *Belief and Probability*. 1976.
105. Kurt H. Wolff, *Surrender and Catch: Experience and Inquiry Today*. Boston Studies in the Philosophy of Science, Volume LI. 1976.
106. Karel Kosík, *Dialectics of the Concrete*. Boston Studies in the Philosophy of Science, Volume LII. 1976.
107. Nelson Goodman, *The Structure of Appearance*. (Third edition.) Boston Studies in the Philosophy of Science, Volume LIII. 1977.
108. Jerzy Giedymin (ed.), *Kazimierz Ajdukiewicz: The Scientific World-Perspective and Other Essays, 1931-1963*. 1978.
109. Robert L. Causey, *Unity of Science*. 1977.
110. Richard E. Grandy, *Advanced Logic for Applications*. 1977.
111. Robert P. McArthur, *Tense Logic*. 1976.
112. Lars Lindahl, *Position and Change. A Study in Law and Logic*. 1977.
113. Raimo Tuomela, *Dispositions*. 1978.
114 Herbert A. Simon, *Models of Discovery and Other Topics in the Methods of Science*. Boston Studies in the Philosophy of Science, Volume LIV. 1977.
115. Roger D. Rosenkrantz, *Inference, Method and Decision*. 1977.
116. Raimo Tuomela, *Human Action and Its Explanation. A Study on the Philosophical Foundations of Psychology*. 1977.
117. Morris Lazerowitz, *The Language of Philosophy. Freud and Wittgenstein*. Boston Studies in the Philosophy of Science, Volume LV. 1977.
118. Stanislaw Leśniewski, *Collected Works* (ed. by S. J. Surma, J. T. J. Srzednicki, and D. I. Barnett, with an annotated bibliography by V. Frederick Rickey). 1982. (Forthcoming.)
119. Jerzy Pelc, *Semiotics in Poland, 1894-1969*. 1978.
120. Ingmar Pörn, *Action Theory and Social Science. Some Formal Models*. 1977.
121. Joseph Margolis, *Persons and Minds. The Prospects of Nonreductive Materialism*. Boston Studies in the Philosophy of Science, Volume LVII. 1977.
122. Jaakko Hintikka, Ilkka Niiniluoto, and Esa Saarinen (eds.), *Essays on Mathematical and Philosophical Logic*. 1978.
123. Theo A. F. Kuipers, *Studies in Inductive Probability and Rational Expectation*. 1978.

124. Esa Saarinen, Risto Hilpinen, Ilkka Niiniluoto, and Merrill Provence Hintikka (eds.), *Essays in Honour of Jaakko Hintikka on the Occasion of His Fiftieth Birthday.* 1978.
125 Gerard Radnitzky and Gunnar Andersson (eds.), *Progress and Rationality in Science.* Boston Studies in the Philosophy of Science, Volume LVIII. 1978.
126. Peter Mittelstaedt, *Quantum Logic.* 1978.
127. Kenneth A. Bowen, *Model Theory for Modal Logic. Kripke Models for Modal Predicate Calculi.* 1978.
128. Howard Alexander Bursen, *Dismantling the Memory Machine. A Philosophical Investigation of Machine Theories of Memory.* 1978.
129. Marx W. Wartofsky, *Models: Representation and the Scientific Understanding.* Boston Studies in the Philosophy of Science, Volume XLVIII. 1979.
130. Don Ihde, *Technics and Praxis. A Philosophy of Technology.* Boston Studies in the Philosophy of Science, Volume XXIV. 1978.
131. Jerzy J. Wiatr (ed.), *Polish Essays in the Methodology of the Social Sciences.* Boston Studies in the Philosophy of Science, Volume XXIX. 1979.
132. Wesley C. Salmon (ed.), *Hans Reichenbach: Logical Empiricist.* 1979.
133. Peter Bieri, Rolf-P. Horstmann, and Lorenz Krüger (eds.), *Transcendental Arguments in Science. Essays in Epistemology.* 1979.
134. Mihailo Marković and Gajo Petrović (eds.), *Praxis. Yugoslav Essays in the Philosophy and Methodology of the Social Sciences.* Boston Studies in the Philosophy of Science, Volume XXXVI. 1979.
135. Ryszard Wójcicki, *Topics in the Formal Methodology of Empirical Sciences.* 1979.
136. Gerard Radnitzky and Gunnar Andersson (eds.), *The Structure and Development of Science.* Boston Studies in the Philosophy of Science, Volume LIX. 1979.
137. Judson Chambers Webb. *Mechanism, Mentalism, and Metamathematics. An Essay on Finitism.* 1980.
138. D. F. Gustafson and B. L. Tapscott (eds.), *Body, Mind, and Method. Essays in Honor of Virgil C. Aldrich.* 1979.
139. Leszek Nowak, *The Structure of Idealization. Towards a Systematic Interpretation of the Marxian Idea of Science.* 1979.
140. Chaim Perelman, *The New Rhetoric and the Humanities. Essays on Rhetoric and Its Applications.* 1979.
141. Wlodzimierz Rabinowicz, *Universalizability. A Study in Morals and Metaphysics.* 1979.
142. Chaim Perelman, *Justice, Law, and Argument. Essays on Moral and Legal Reasoning.* 1980.
143. Stig Kanger and Sven Öhman (eds.), *Philosophy and Grammar. Papers on the Occasion of the Quincentennial of Uppsala University.* 1981.
144. Tadeusz Pawlowski, *Concept Formation in the Humanities and the Social Sciences.* 1980.
145. Jaakko Hintikka, David Gruender, and Evandro Agazzi (eds.), *Theory Change, Ancient Axiomatics, and Galileo's Methodology. Proceedings of the 1978 Pisa Conference on the History and Philosophy of Science, Volume I.* 1981.
146. Jaakko Hintikka, David Gruender, and Evandro Agazzi (eds.), *Probabilistic Thinking, Thermodynamics, and the Interaction of the History and Philosophy of*

Science. Proceedings of the 1978 Pisa Conference on the History and Philosophy of Science, Volume II. 1981.

147. Uwe Mönnich (ed.), *Aspects of Philosophical Logic. Some Logical Forays into Central Notions of Linguistics and Philosophy.* 1981.
148. Dov M. Gabbay, *Semantical Investigations in Heyting's Intuitionistic Logic.* 1981.
149. Evandro Agazzi (ed.), *Modern Logic – A Survey. Historical, Philosophical, and Mathematical Aspects of Modern Logic and its Applications.* 1981.
150. A. F. Parker-Rhodes, *The Theory of Indistinguishables. A Search for Explanatory Principles below the Level of Physics.* 1981.
151. J. C. Pitt, *Pictures, Images, and Conceptual Change. An Analysis of Wilfrid Sellars' Philosophy of Science.* 1981.
152. R. Hilpinen (ed.), *New Studies in Deontic Logic. Norms, Actions, and the Foundations of Ethics.* 1981.
153. C. Dilworth, *Scientific Progress. A Study Concerning the Nature of the Relation Between Successive Scientific Theories.* 1981.
154. D. W. Smith and R. McIntyre, *Husserl and Intentionality. A Study of Mind, Meaning, and Language.* 1982.
155. R. J. Nelson, *The Logic of Mind.* 1982.
156. J. F. A. K. van Benthem, *The Logic of Time. A Model-Theoretic Investigation into the Varieties of Temporal Ontology, and Temporal Discourse.* 1982.
157. R. Swinburne (ed.), *Space, Time and Causality.* 1982.
158. R. D. Rozenkrantz, *E. T. Jaynes: Papers on Probability, Statistics and Statistical Physics.* 1983.
159. T. Chapman, *Time: A Philosophical Analysis.* 1982.
160. E. N. Zalta, *Abstract Objects. An Introduction to Axiomatic Metaphysics.* 1983,
161. S. Harding and M. B. Hintikka (eds.), *Discovering Reality. Feminist Perspectives on Epistemology, Metaphysics, Methodology, and Philosophy of Science.* 1983.
162. M. A. Stewart (ed.), *Law, Morality and Rights.* 1983.
163. D. Mayr and G. Süssmann (eds.), *Space, Time, and Mechanics. Basic Structure of a Physical Theory.* 1983.
164. D. Gabbay and F. Guenthner (eds.), *Handbook of Philosophical Logic.* Vol. I. 1983.
165. D. Gabbay and F. Guenthner (eds.), *Handbook of Philosophical Logic.* Vol. II, forthcoming.
166. D. Gabbay and F. Guenthner (eds.), *Handbook of Philosophical Logic.* Vol. III, forthcoming.
167. D. Gabbay and F. Guenthner (eds.), *Handbook of Philosophical Logic.* Vol.IV, forthcoming.
168. Andrew, J. I. Jones, *Communication and Meaning.* 1983.
169. Melvin Fitting, *Proof Methods for Modal and Intuitionistic Logics.* 1983.